10 0152204 1

UNIVERSITY OF NOTTINGHAM

WITHDRAWN FROM THE LIBRARY

DATE DUE FOR RETURN

UNIVERSITY LIBRARY

31 JAN 2006

SEM GGL 07

UNIVERSITY LIBRA

07 JUN 2006

UNIVERSITY LIBRARY

MANUFACTURING PLANT LAYOUT

MANUFACTURING PLANT LAYOUT

Fundamentals and fine points of optimum facility design

By Edward J. Phillips, P.E.

Society of Manufacturing Engineers
One SME Drive, P.O. Box 930
Dearborn, Michigan 48121

Copyright © 1997 by Edward J. Phillips

98765432

All rights reserved, including those of translation. This book, or parts thereof, may not be reproduced in any form or by any means, including photocopying, recording, or microfilming, or by any information storage and retrieval system, without permission in writing of the copyright owner.

No liability is assumed by the publisher with respect to the use of information contained herein. While every precaution has been taken in the preparation of this book, the publisher assumes no responsibility for errors or omissions. Publication of any data in this book does not constitute a recommendation or endorsement of any patent, proprietary right, or product that may be involved.

Library of Congress Catalog Card Number: 97-065261
International Standard Book Number: 0-87263-484-1

Additional copies of this book may be obtained by contacting:

Society of Manufacturing Engineers
Customer Service
One SME Drive
Dearborn, Michigan 48121
1-800-733-4763

SME staff who participated in producing this book:

Donald A. Peterson, Senior Editor
Rosemary K. Csizmadia, Production Team Leader
Dorothy M. Wylo, Production Assistant
Jennifer L. Courter, Editorial Intern
Frances M. Kania, Production Assistant

Karen M. Wilhelm, Manager, Book Publishing

Cover design: Judy D. Munro, Manager, Graphic Services

Printed in the United States of America

1001522041

PREFACE

Over the last quarter century, manufacturers and software developers have designed parts and product costing systems in great detail. Much of the technology of present-day costing techniques stems from basic rudimentary manufacturing engineering and industrial engineering studies of detail manufacturing processes. Most of the technology is now minicomputer and personal-computer based, having no absolute need for a large mainframe computer. Moreover, excellent costing systems are now widely available around the globe. Because of these advancements in computer capability and software, most manufacturers who have been in business for more than 5 or 10 years believe they have a relatively good grasp of cost within their factories. But in fact, do they?

Many indirect costs affect a company's performance. A poor plant layout can have a particularly adverse effect on performance and costs. In my experience, very few manufacturers recognize the deleterious effects a poor plant layout can have on employee morale, productivity, and total factory costs. Many companies will suffer a poor plant layout for years and even decades before either competition or education opens their eyes. Plant layout problems are generally latent or hidden, which often makes them difficult — if not impossible — to see by many mid- and upper-level managers.

Many manufacturers have a good grasp of the *direct* labor and materials costs of their manufacturing processes; most, however, have little knowledge of their *indirect* materials handling costs. For purposes of this book, indirect costs are described as those costs associated with the plant layout. A major cost in that regard includes the costs of transporting materials throughout the plant and storing them within it.

Although in-plant materials handling and plant layout changes may not have the "glamour" or total savings potential that a significant change in processes or manufacturing methods might have, the often-overlooked costs of layout and handling can be quite high. In fact, hidden indirect cost problems can be far too high to ignore. Usually, however, it does take an eye-opening event or radical change before a poor plant layout is even recognized. When concentrating on "tweaking" the manufacturing processes to an absolute optimum, one can easily "miss the forest for the trees." In other words, in-plant materials handling inefficiencies, as a result of a poor plant layout, may be easily overlooked when one is occupied with fixing day-to-day problems in the detail manufacturing processes.

In those cases where a poor plant layout may be very obvious to plant management, company officials may not have the time, patience, or training to deal with it. Few people (visionaries?) are able to visualize and realize that reviewing the logistics of the total plant may offer tremendous opportunities for improvement. As a result, improvements are generally done on an ad hoc, piecemeal basis.

The objective of this book is to provide the reader with sound methods for developing an optimum plant layout. In writing it, I have followed a nonacademic, practical approach used extensively by competent industrial consultants. The subject matter does not dwell on theoretical or scholarly approaches to plant layout. Instead the book focuses on the methods I have used successfully to develop real-world solutions to real-world industrial problems. I have tried to be as objective, graphical, and quantitative as possible. Although one can never totally eliminate subjectivity in the layout process, I have attempted to reduce the impact of subjective data to the greatest extent.

The book is based largely on my experience and consulting work as managing director of The Sims Consulting Group, Inc. The work draws very heavily from both the public and private seminars I have conducted in the U.S. and Asia over the last decade. More than a thousand engineers and manufacturing managers have participated in these programs and much of this work is based on their experience. The book also draws heavily on my early experience as a consultant working with Richard Muther. Dick Muther was and still is a true pioneer in developing a systematic, engineered approach to solving facilities-related problems. Almost all recognized work in plant layout and facilities planning can be traced back to Dick's original pioneering efforts as discussed in his book *Systematic Layout Planning**. He introduced me to a consulting career and guided my original efforts in consulting assignments, conducting seminars, and publishing professional papers. Much of this book builds on Muther's original work. I also owe a debt of gratitude to another industrial engineering pioneer and mentor, E. Ralph Sims, Jr., who prodded and cajoled me for more than 5 years to write this book.

ACKNOWLEDGMENTS

I would like to express my deep appreciation to several people and organizations that helped bring this book to fruition: Richard Koons of The Sims Consulting Group, Inc., who helped in the preparation of many of the graphics, Dave Sly of Cimtechnologies Corporation, who provided much of the material on software tools for static analyses of macro flows and costs, and Jerry Hoskins of Manufacturing Engineering, Inc., who provided the information on computer-based "Dynamic Model Building." Many thanks also to Kathleen Phillips, my wife, for her ongoing moral support and putting up with many "lost" evenings and weekends as the book was being written.

I would appreciate hearing from the book's readers. Your comments, concerns, and suggestions will give guidance to a possible future edition. You may e-mail comments directly to me at sims968@aol.com, or U.S. mail to:

P. O. Box 2584
Lancaster, Ohio 43130.

Or, you may contact the publisher, the Society of Manufacturing Engineers (SME) at:

SME Book Publishing Department
Attn: Don Peterson
One SME Drive
P.O. Box 930
Dearborn, Michigan 48121-0930

Ed Phillips
January, 1997

---

*Muther, Richard; *Systematic Layout Planning* (SLP); 485 p.; Management and Industrial Resource Publications; 1987.

CONTENTS

Preface v

    Acknowledgments vi

1. Introduction 1

    Competitive business challenges.. 1

    The continuous improvement doctrine.. 2

    Cycle times and materials handling—

        the difference that makes "the difference" in a global economy 3

    The key fundamentals in manufacturing.. 5

2. An Overview of the Major Planning Phases 7

    The five phases of developing manufacturing plant layouts.................... 7

    Manufacturing process planning 8

    Equipment and systems planning 9

    Space planning... 10

    Site selection 14

    Evaluating alternative sites 16

    Overall site planning 18

    Block layout planning... 19

    Schematic utilities/energy planning... 24

    Detail layout planning 25

    Implementation planning 27

3. Where to Begin 37

    The need for change 37

    People skills and leadership 39

    The computer—an invaluable project management tool......................... 39

    Setting the path to success—defining the attributes................................ 40

    A typical layout planning approach 41

    A prelude to the data gathering process 41

    Planning conventions and standard symbols 42

4. **Basic Data Needs** **43**
   Data review and analyses—four major facets 43
   Categorizing data needs 43
   Fundamental data 45
   Issues and problems in data collection 46
   Product versus quantity 50
   Process versus product 54
   Scheduled push versus demand pull 59

5. **Materials Handling Analysis** **63**
   Bulk versus unit materials handling 63
   The basic questions for work simplification 63
   Large versus small unit loads 65
   The effects of unit load and container configuration 67
   Dual manufacturing and shipping containers—a special case 68
   Typical manufacturing plant flow patterns 69
   The operations process (flow) chart 71
   The importance and opportunities of production line and cell balances 79
   Equivalent unit load analysis 85
   The from-to chart 88
   Basic factors affecting all moves 92

6. **Calculating Space Requirements** **95**
   General considerations 95
   Use of gross business ratios 95
   Space balance for long-term projections 98
   Site saturation/master planning method 101
   Equipment utilization considerations 105
   Adjusting today's needs 106
   Detail determination/calculation of space needs 109
   Planning storage areas within the plant 109
   Ratio trend and projections 125
   Pitfalls and realities of space projections 127

7. **Relationship/Affinity Analysis** **129**
   The engineered planning approach 129
   The three *As* of plant layout 133
   Methods of establishing relationships between activities 134
   Getting the logic approvals 150
   The affinity, proximity, relationship, bubble diagram—take your choice 151
   Typical computer-aided approaches 157
   Benefits and pitfalls of using relationship diagrams 160

8. **Establishing Relationship Diagrams for Existing Plants** **163**
   New plant versus expansion 163
   Pre-diagramming: practical considerations/attitudes 163
   Existing constraints and monuments 166

9. **Developing the Spatial Relationship Diagram** **169**
General considerations 169
Diagramming for a "greenfield" site 169
Diagramming for an existing plant or plant expansion 171
Important physical considerations 171
Planning natural aisle patterns 172

10. **Developing Alternative Layout Configurations** **175**
Keeping your eyes on the logic and the attributes 175
Expansion flexibility considerations 176
Intradepartment/cell detail layout considerations 180
Ergonomics/human factors 186
Budget limitations 190

11. **Manufacturing Cells** **197**
Cell history 197
Group technology 199
The three major *P* factors in cell planning 201
Reducing nonvalue-added operations 203
Cell planning procedure 204
Detail cell planning 210
Balancing labor within cells and on assembly lines 219

12. **Multifloor, Multisite Space Allocations** **225**
Another use for relationship analyses 225
Computer allocation systems 226

13. **Computer-based Tools** **231**
Computer analyses of macro flows and costs 231
Dynamic computer simulation 234
Dynamic model building 236
Second example 240

14. **Evaluating Alternatives** **245**
Tangible and intangible decision factors 245
Setting the framework for team decision-making — using weighted factor evaluation 245

15. **Closing Notes on Relationships and Materials Handling Equipment Choices** **249**
Benefits of a well-designed materials handling system 249

Appendix **253**

Index **255**

1

INTRODUCTION

COMPETITIVE BUSINESS CHALLENGES

Have you ever wondered why some companies' products always seem to have a competitive cost advantage over others? Have you ever wondered what those companies might be doing differently from what you are doing?

One thing successful companies almost always have in common is a streamlined organization and a _streamlined plant layout_. There have been dozens, if not hundreds, of books written on the subject of streamlining _organizations_. This book is dedicated to helping you and your organization achieve competitive success through streamlining your _plant layout_ and the operations housed within those layouts.

Savings in material handling costs, shop time, work-in-process (WIP) inventories, and scrap from excessive material handling _can_ and _do_ make a huge difference. After all, what tasks make up the largest components of factory labor and the largest components of the cost of materials purchased and used in factory processes? They are the handling, storage, and movement of materials in our plants or in the plants of our suppliers. Some accountants would argue that materials cost is a bigger portion of the total cost pie than materials handling. That may be true from a purchasing perspective, but if you could "see" inside your supplier's plants, you would note that a large portion of that supplier's total costs is normally associated with internal layout and materials handling. Those vendor inefficiencies are passed along to you as part of your purchased materials cost.

The competitive advantage that can be achieved with a well-planned and implemented plant layout cannot be ignored. This facet of the manufacturing enterprise is one of the latent factors of competitive success.

It is no secret that today's world is much more competitive than the world in which our parents and grandparents worked. In the first half of the 20th century, much of the industrialized world held semi-isolationist attitudes. Most of the manufactured products that could be purchased by consumers in the western industrialized countries were made within the boundaries of their own country. This situation generally held true through the post-World War II period of the late 1940s and into the early 1960s. Since that time, there has been a dramatic shift in the worldwide manufacturing base.

Today, as we approach the 21st century, it is extremely difficult to shop at a marketplace and find an abundance of manufactured products that are made in one's own country. To demonstrate this point, if you live in an "open-market" country, such as the United States, the United Kingdom, or one of the western European nations, shop in any major metropolitan area and find a dozen or so different manufactured products made totally in your own country. That may sound easy enough, but it is not. This is particularly true for products that sell in a price category between the value of 1 to 4 hours of a given country's average factory wage.

Changes and fundamental shifts in the manufacturing base have brought competitive business challenges to all of us. The world continues to become effectively smaller with many more competitors than heretofore. The advances in communications technology, e.g., instantaneous e-mail, wireless remote computer access, the Internet, etc., are shrinking the world even more. These changes will absolutely force each manufacturing enterprise to become much more efficient on a continuing basis. It is not simply a question of higher or lower factory costs, it is truly a question of survival of the fittest. Each component of cost and quality needs to be continually improved or our competitive edge will be lost.

In the past, a new factory layout or plant rearrangement was completed by the company's draftsperson who may have had very little manufacturing process knowledge. Conventional wisdom of the day was that great expertise associated with plant layouts did not exist. Any special problems that developed could be overcome by adding an additional fork truck or length of conveyor. Even today, there are many uninformed executives who still do not realize the ongoing negative impacts on people and costs that result from a bad plant layout. Just a sampling are:

- High materials handling costs,
- Cycle and lead time delays,
- High work-in-process inventories,
- Lower than optimum quality,
- Product or parts damage,
- Safety and morale problems,
- Poor equipment utilization (both fixed and mobile),
- Congested aisles,
- Wasted floor space.

Companies that want to compete in a world-class, global economy can no longer ignore the ongoing "costs" of a poor plant layout. Once the layout and equipment have been put in place, it may be too late to justify changes. We need to constantly review and improve our manufacturing methods and the plant layouts that support those methods.

THE CONTINUOUS IMPROVEMENT DOCTRINE

Why do so many companies never even come close to achieving an optimum plant layout and streamlined production and product flow? Why

do they run the risk of losing a potential cost advantage to competitors?

Can management not see the hidden cost impact of a poor plant layout on employee morale and respect for management? Is it a lack of management commitment or is it instead a lack of knowledge?

The competitive challenges faced by manufacturing organizations are not discrete events. One cannot simply go through an annual budget exercise and commit to this or that particular project or plant layout. We all know (or at least we should know) the fallacies of short-term management approaches. Consider this example of one of the problems the "discrete event" approach fosters.

Suppose you are designing an automated manufacturing cell. In your design for the total system, you develop an operations and plant layout plan to completely automate a series of sequential operations that are currently being performed manually. You get preliminary plan approval and submit the budgetary appropriation request for the total system to upper-level management. Everything is fine until this point, but now management is balking at implementing the total system all at once. Investment funds are scarce and there are other projects that also have high priorities. (You may also be employed by a company that takes a short-term view rather than a strategic view of manufacturing.) A "piecemeal" approach may be decided upon with the total project never being fully completed and implemented. In that event, the company will be forced to pay the higher materials handling costs (as well as the other "costs" discussed previously) associated with the less-than-optimum layout. These higher costs are *usually incurred for many years!*

What happened? When executive-level management is faced with tough budgetary decisions, they will sometimes elect to approve only those portions of the "total system" plan that will offer the highest short-term return on investment (ROI).

With tight budgetary constraints, an attempt will be made to put off the smaller ROI portions of the project. Most total "manufacturing systems" plans have capital investment components that offer a wide swing in ROI, some being very small but absolutely essential to the whole. Once the "cream of the crop" portions of the project are approved, it is difficult, if not impossible, to later gain an approval of those capital portions of the project with a small ROI as standalone entities. Unfortunately, the development of optimum plant layouts is normally tied to these discrete capital improvement projects.

What does this mean to us? It means a company with that kind of short-term, discrete-event philosophy will *never* come close to having the optimum system or optimum plant layout. By definition, it will never be able to justify, on a standalone basis, those portions of the project that were absolutely essential to achieving maximum productivity and minimum materials handling costs. However, someday the company is likely to face a competitor *who does opt to complete the whole system*. It is even more likely that the company will face a competitor who *views the development of optimum plant layouts as a continuous process* and not a discrete event tied to a capital project. From that point on, the company with short-term thinking will continuously be at a disadvantage.

The continuous improvement doctrine (CID), as the name implies, incorporates a philosophy of ongoing improvement. A company with a true CID philosophy would have absolutely committed itself to the "total" system approach even though funding for some of the smaller ROI items may have been delayed. But, even though funding may have been delayed, *all* of the project work would eventually have been implemented. The plant layout would be developed for the total system, even if the company needed to absorb some higher materials handling costs in the short term.

This continuing problem of attempting to incrementally implement optimum manufacturing plant layouts forces many companies to literally suffer for years with a poor plant layout because they do not believe they can economically justify making improvements. Yet they bitterly complain about unfair competition when a more farsighted competitor with a CID vision *does* implement layout improvements on an ongoing basis. This competitor will naturally have lower factory and product costs.

A continuous, ongoing improvement plan for production processes, materials handling, and plant layout is essential to achieving competitive advantage. The plan needs to be developed and maintained on an ongoing basis.

CYCLE TIMES AND MATERIALS HANDLING — THE DIFFERENCE THAT MAKES "THE DIFFERENCE" IN A GLOBAL ECONOMY

Cycle Times

How often have you heard a manufacturing person say "the smaller the lot size, the higher the materials handling and setup costs," or "the faster the assembly line conveyor speed, the higher the production output?" Adages die slowly.

I remember visiting a brand-name appliance manufacturer's plant in the midwest U.S. several years ago. I also vividly remember almost ripping my pants to shreds on the WIP inventory of sheet-metal parts that were literally strewn throughout the entire plant. There was not an aisle in that plant that did not have parts stored in it. There was also storage on an overhead monorail conveyor. When I asked the plant manager what rationale he used to determine lot sizes for materials handling, he said the handling made no difference! He claimed his lot size was based on minimizing setup time on his steel coil line, punch presses, and press brakes. He felt he had a very efficient operation. Why? Because he only needed to change part setups when each 40,000-pound coil of steel ran out! He perceived his goal to be that of minimizing setup times even though he was building (and his company was financing) a ridiculous amount of unneeded inventory.

The same plant had a final assembly line producing around 1,400 completed units of various appliance models each work shift. The units were placed approximately 30 inches away from one another on the moving line and the line speed was set so that a finished unit came off the line about every 18 seconds. The company was attempting to improve productivity by implementing a line "speed-up" program, but management was having problems with the union and a number of assembly line employees when it tried to crank up the line speed. The line assemblers felt they were working much too hard prior to the speed-up, let alone at a ramped-up pace. As it turned out, *they were in fact working much too hard.* The initial line speed forced most assemblers to move a step forward to complete their tasks to keep pace with the line speed. Then they had to take a step back to get to the starting point for the next unit at their station. This process was repeated for each unit: it was extremely tiring for many operators. A simple change was implemented which allowed the line to be slowed down and which, paradoxically, improved productivity dramatically.

It was not easy to change the plant culture that had been in place for years. Company management's initial reaction to our request to slow the line down was a shout of: "Hogwash —

you don't know what you are talking about. How is it possible to slow the line down and get more production out?" Actually, it was very possible and also very simple. We reduced the space between units to 12 inches instead of 30 and slowed the line speed down from 200 inches per minute (ipm) to 168 (see Figure 1-1).

This gave us a new cycle time of 15 seconds, a significant improvement over the previous 18-second cycle. Shift production (7 hours) increased to 1,680 units as opposed to the previous 1,400, yielding a 20% production increase. The employees also were much happier. The slower line speed helped eliminate the unneeded paces they previously had to take. Keep in mind, the actual assembly labor content at each position did not change—the lost time due to unnecessary stretching and pacing was eliminated. (We don't mean to imply that this was an overly

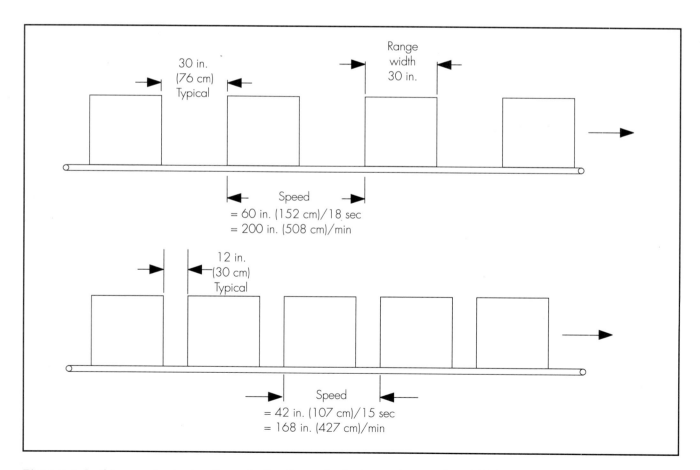

Figure 1-1. *In this example, slowing the production line and reducing product spacing yielded a 20% boost in productivity.*

simple correction; some marginal positions *did* have to be rearranged and rebalanced.) Most of the assembly-line people could now perform their tasks almost sitting down. There was much less stretching and pacing. They perceived themselves to be working "smarter, not harder."

The real point to be made here is that, prior to this change, a mindset had developed over the years that an 18-second cycle was the optimum. This cycle time had been in effect for the previous 5 years. The company's costs continued to rise until they had no choice but to *do something*. For its part, the company could only think of speeding the line pace. A more progressive company, with a continuous improvement mindset, would have been "tweaking" and improving the line balance and productivity of the assembly line all the time. Both the employees and management should have developed an expectation of at least semi-constant change.

Materials Handling

I have also seen companies try to rationalize plant layouts and materials handling plans that include very long employee walks or long industrial truck movements. In one fabrication shop, at the end of a particular process, employees manually moved push carts of materials some 480 feet to the next operation. Mechanization of the effort was ruled out as too expensive. Management tried to justify the layout using a highly detailed industrial engineering analysis. (The company was considering the purchase of a very long and narrow building at a "good" price.) The industrial engineering analysis showed that the round trip would take just over 6 minutes to complete. This was deemed acceptable to management. What the analysis did not show was that an employee making the trip would stop midway to chat with another employee for 4 or 5 minutes. He or she would then stop *at least* one time again on the return trip to chat with one or two other employees for 4 or 5 minutes, and so on. When the potential lost production time of these employees was taken into account, lost productivity was closer to 30 minutes. When the *ripple effects* of other employees who could be held up because one of their associates stopped for a "chat" was taken into account, that original 6-minute walk exploded into

40 or 50 minutes of potential lost productivity. When this was extended against the total number of annual trips required on all three shifts, the potential total loss in productivity was far more than the 6 minutes management had expected. It was finally recognized that the short-term penalty from such a layout was not acceptable and the long-term penalty would have been intolerable. The company decided against purchasing the "long" building.

Some may argue that perhaps this was a case of employee discipline and not a layout or materials handling problem. Or, that perhaps a conveyor or automatic guided vehicle could have been used. Since all problems in layout and materials handling involve choices and alternatives, you should draw your own conclusions. However, all decisions need to be made in the context of the business bottom line, i.e., competition. What would our global competitors do in a similar situation? What about potential new competitors? Would a competitive company intentionally purchase a suboptimum facility if it had a choice?

We truly are in a global economy. For industrialized nations to compete in manufacturing against one another and also against Third World labor costs, there must be a significant difference in materials handling and manufacturing methods. One cannot simply accept the status quo because the company for which one works has a "good" year. Everyone should constantly question materials handling methods and manufacturing methods. A policy of continuing to perform the same way, year after year, because "we have always done it that way" is doomed to fail.

THE KEY FUNDAMENTALS IN MANUFACTURING

Our customers expect four fundamentals from our manufacturing business, all related to our customer's expectations.

1. Product quality—both perceived and actual.
2. Delivery as scheduled/requested—not before and not after.
3. Flexibility to handle change and service—production schedule, product configuration, customer deliveries, etc.
4. Low prices.

Product quality may be affected by a great many items, some of which are beyond the scope of this text. We can state that, on balance, manufacturing cell (single-part or very small batch flow), product-focused plant layouts tend to have higher-quality output when compared with functionally-segregated, large-scale, batch-process-oriented layouts. This is certainly not true in all cases, but it usually holds for some portion of products in most companies engaged in light manufacturing.

Focused business units with cellular manufacturing can also facilitate lower cycle/lead times through the shop, which can improve delivery performance. The focused business unit philosophy has the added benefit of increasing customer confidence. However, in a mixed factory environment, with several business units competing for support services, there needs to be balance. That balance revolves around the issue of which functions should be centralized and serve all units and which should be totally included within each factory business unit or cell.

Cellular manufacturing, coupled with a *demand pull* scheduling system, can also help facilitate large reductions in work-in-process (WIP) inventories or almost eliminate them. Reduction in WIP inventories can impact the company's bottom-line performance significantly. However, it is not the intent in this introduction to give the impression that this book is devoted totally to cellular manufacturing. There are both good and bad attributes associated with centralized batch manufacturing environments and decentralized cellular manufacturing approaches. There are also several other influencing factors and positives and negatives of both approaches that we discuss in a later section.

It is no secret that to keep our product prices low, we need to consistently minimize overhead and factory costs. Two major elements of the direct and indirect factory labor costs equation are the way we handle materials and the way we lay out our plant. These two facets of the manufacturing plant are the principal focus of this book.

2

AN OVERVIEW OF THE MAJOR PLANNING PHASES

THE FIVE PHASES OF DEVELOPING MANUFACTURING PLANT LAYOUTS

Figure 2-1 shows how the major planning phases of a new plant project or a major expansion planning overlap. The five phases through which any layout project passes are:

1. *Needs Analysis.* This is a determination of what is actually required to correct problems or meet new challenges. Usually this is a lengthy process which overlaps at least two other phases (and sometimes more, as the company's needs may change throughout the project). It basically involves developing the overall production needs and area requirements for the new site or an internal expansion. This phase also involves the development of strategic goals regarding the manufacturing techniques and approaches to be employed (e.g., cellular versus traditional batch manufacturing).

2. *Location Analysis.* In this phase, determination of where the location is to be laid out is made. Location analysis may involve a new site, but more often it involves redesigning the present location, rearranging several locations within the present plant, or arranging a newly acquired building or some other available space. Sometimes the layout planner may not be directly involved in this decision, particularly if a new site is being considered. However, it is highly recommended that the layout planner be involved in the evaluation of potential sites.

3. *Block Layout.* This phase determines the basic flow patterns and major individual area allocations. It provides the general size and configuration of each major area, and proximity relationships, affinities, and major material flows between these areas. Several block layout alternatives are usually developed. Main aisles and aisle patterns should be included in this phase.

4. *Detail Layout.* Determination of the specific location of each piece of machinery, equipment, or physical feature within the layout, including utilities and services is made during the detail layout phase. This detail layout may be a series of drawings, several drawing layers in a CAD system, or a three-dimensional replica board on which replicas of all machinery, equipment, and physical features are placed.

5. *Installation.* In this phase, installation instructions and time line schedules are developed, and the various approvals, appropriations, and permits needed to perform the actual installation are secured. Typically, a major facet of this phase is scheduling and implementing the rearrangement or move with the least amount of disruption to ongoing production operations.

It is important to recognize that it is far better to work from the top down than from the bottom up. Less backtracking and rework and better results will be obtained if one moves progressively from gross requirements to detail

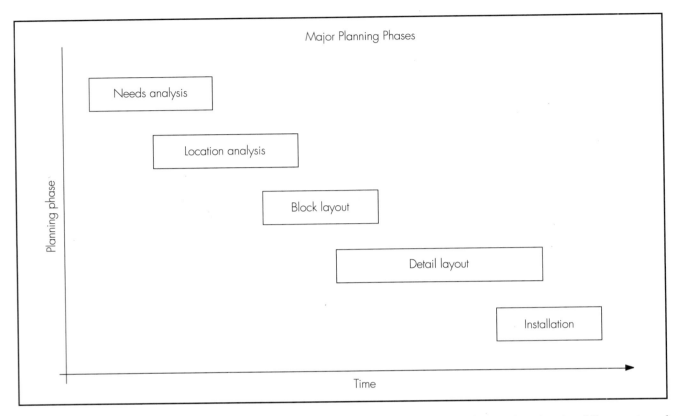

Figure 2-1. *From needs analysis through the installation process, effective new plant and expansion planning follows a stepped approach of overlapping phases.*

requirements. For example, it is better to select a site before completing block layout alternatives, and it is better to develop block layouts before completing detail layouts.

However, it is also important to recognize the overlapping nature of the phases. Even though the phases appear to be theoretically sequential, it would be a mistake to attempt to perform them *totally* sequentially without an overlap. For example, it would be foolhardy to attempt a block layout without some knowledge of the detail sizes and configurations of equipment to be installed within the block. It is insufficient, for example, to know that a paint line to be installed takes up an area of 17,200-plus ft² (1,600 m²). The planner could incorrectly assume a square block area layout of 131 ft (40 m) wide by 131 ft long when the actual system may be 65 ft (20 m) wide by 263 ft (80 m) long. An ideal block layout would be worthless if the planned

equipment does not fit properly in the allocated blocks of space.

MANUFACTURING PROCESS PLANNING

Before we can begin to structure a task schedule for the development of plant layouts, materials handling plans, and an overall facilities configuration, we need to determine what manufacturing processes and philosophies will be used within the plant. The statement assumes that we already have some knowledge of unit quantity requirements. This knowledge should encompass production and shipping estimates for at least the next 4 to 5 years. Some companies will extend the horizon of projections to 10 or more years using long-term analysis techniques. High-tech companies may not have the luxury of realistic long-term projections and must plan for the utmost in flexibility. Obviously, the farther out anyone's projections go, the less accurate the projections become.

For major rearrangements and new plants, some of the initial basic questions to be considered are:

- Will we be using continuous, flow-through processes or batch-oriented, job-shop types of processes?
- Will we be considering a segregated "process" type of functional layout or a segregated "product" type of cell layout? Will there be a mix of the two? Should several alternatives be developed?
- For large products, will we build the units in place or will we transport the units through the various stages of production?
- Will we be considering a multifloor processing installation? If so, how many levels?
- What cycle times are we targeting for? At what equipment and labor utilization levels?
- What production and production support operations will be centralized or decentralized?
- Will we be operating on a just-in-time (JIT) basis? How much staging (in time units) should be allowed for materials—30 minutes, 1 hour, 1 shift, 1 day, etc.?
- What amount of process downtime should we allow for? What does the expected distribution curve for downtime look like?
- What is the company's philosophy regarding buffer storage within a cell, on an assembly line, or between processes? If an operation within a cell or assembly line is down, do we stop the entire cell or do we allow the front end of the operations to continue to build to a buffer storage area until repairs are made? If we continue to build, how large a buffer area will be required?
- How many daily, weekly, or monthly work shifts will be employed?
- How will scrap, chips, etc. be handled and accumulated?
- What immediate methods improvements should be planned? What 1- to 2-year improvements should we consider? What longer-term process and methods improvements should we take into account?

In considering these issues (and the following overview items), the planner should not be intimidated into believing that all of the information has to be available before he or she can start the project. It is highly unlikely that all of the questions will have been answered until much further along in the layout and configuration process. Although much of the manufacturing process planning is predetermined, several layout iterations may be required before we can determine what alternative processes should be selected. Answers to some strategic questions will emerge through refinement as the layout progresses. In later sections of the book, we will discuss some of the data gathering requirements of both processes and materials handling.

EQUIPMENT AND SYSTEMS PLANNING

Equipment and systems planning *must* be integrated with manufacturing process planning and layout planning. Similar to manufacturing process planning, before we can begin our facility configuration and layout process, we need to determine the equipment alternatives to be considered or excluded and the *systems philosophies* to be used within the plant. Presumably, we will be working on the development of, or will have already tentatively determined, the manufacturing process equipment to be used in the plant. For our purposes here, the words *equipment* and *systems* relate to materials handling equipment and handling equipment systems.

Some of the initial fundamentals to be considered are:

- Will materials handling be performed by direct or indirect labor? Is there sufficient time within equipment operating cycles for the operator (direct labor) to do his or her own materials handling, or do we want a separate *crew* of indirect labor to perform those tasks? Will we develop different plant layouts before making that determination?
- Have we established a general mode of material handling configuration, i.e., will we be using pallets, totes, baskets, tubs, wheeled carts/containers, or other special containers? Will we use a combination of these? Can we establish a modular, nestable type of handling unit that can accommodate most of our materials and be easily transported? (See Figure 2-2.)

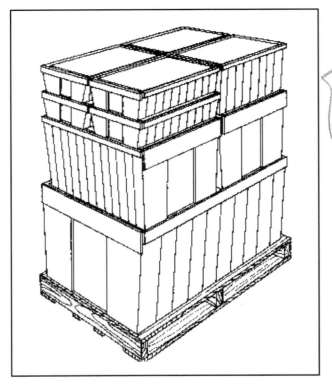

Figure 2-2. Modular, nestable containers are an efficient mode of materials handling and transport.

- For horizontal inplant transport, will we use forklift trucks, "mule-train" pull carts, automatic guided vehicles, manual pushcarts, pallet jacks, conveyors, monorails, overhead cranes, vacuum/pressure tube systems, etc.? Will we use a combination of these? Much of this needs to be determined during the layout process. Has any mode of transport been ruled out?
- Will we be considering vertical transport devices for a multifloor or mezzanine processing arrangement? If so, will we be using power columns, gravity cascades, belt conveyors, spiral chutes, vertical lifts, man-lifts, freight elevators, stacked horizontal carousels with crane pickoffs, automatic storage and retrieval systems, vacuum/pressure tube systems, cranes, etc.?
- What special overhead or supporting load requirements are envisioned? For example, will an overhead crane strategy be used? If so, how many underroof crane bays will we

need? What structural requirements will be needed to support the loads?
- Will we track materials through the plant automatically? Will real-time reporting of materials and process status be employed? If so, what methods will be used — electromechanical sensors, radio frequency, laser scanning of labels or targets, or a combination of these?
- Where will the "detrashing" or depackaging of incoming materials occur — at the receiving dock, within a staging/storage area, at the point of usage? How will the dunnage be handled or conveyed?
- If we plan to use handling containers, how will the empty containers be cycled back to their process starting point? Will this be a people-dependent operation or a more mechanized system? (Generally, people-dependent systems cause "container starvation" at the front end of a process unless the starting and ending points are very close to one another. Container starvation results in capacity losses that can *never* be made up.)

Unlike manufacturing process planning where most processes are predetermined, materials handling equipment systems are almost always designed simultaneously with the plant layout. The marriage of plant layout and materials handling can be likened to good human marriages — compatibility, communication, and consistency will stand the tests of quality and longevity.

SPACE PLANNING

Space planning for new plants and large expansion projects should be a joint effort between architects and experienced industrial and manufacturing engineers.

Most industrial facilities planners believe space planning is the most unscientific and noncreative part of the layout process. Architects, in particular, are very good at what they do. They are extremely creative at space planning and layout development for offices and people-oriented environments. But the great majority of architects are like fish out of water in industrial settings. Architects can certainly design the facility, but most will try to have their customers/clients develop internal factory space needs, processing needs, and the relationships between functional activities based on materials handling analyses.

Conversely, with the proper training, most manufacturing and industrial engineers are quite comfortable and good at developing process needs and the functional relationships between activities based on materials flow. They are also good at developing space needs in factory or warehouse settings. But most would also admit they fall far short of optimum competency in designing office space environments. We are not referring here to simple office layouts with desks and chairs; most manufacturing and industrial engineers can handle that. It is the overall aesthetics—aura if you will—and space attributes that are usually most troublesome to engineers but are the architect's specialty.

In the course of working on more than 100 facilities plans, I have not met more than two individuals who are good at both tasks. Space planning, therefore, should be a joint team effort between architects and experienced industrial and manufacturing engineers.

At this point, however, we need to sound a warning when planning new manufacturing facilities and major expansions. The most common error made by industrial management today is to have the site selection and broad-based architectural "shell" design started (or worse yet, completed) *before* the industrial and manufacturing engineers have started development of alternative facility block layouts. This is, without a doubt, one of the costliest short- and long-term errors a management team can make. It is akin to putting the cart before the horse but with far worse effects. It is understandable that the management team would like to see an artist's rendering of a new plant before approaching the board of directors with a proposal. But the team really does have to do its homework first.

The short- and long-term productivity and materials handling effectiveness of the manufacturing plant is determined by what goes on *inside* the shell, not what the shell looks like from the outside. Invariably, putting architectural development before engineering development places severe constraints on internal manufacturing productivity. It is far better to have a relatively good head start on determining the

internal layout requirements before starting any architectural work or completing the site selection process. It is interesting to note that experienced boards of directors (who have seen or lived with the results of this error) will frequently ask the management team to see their preliminary block layout plans along with the architectural renderings during the initial board presentation. The directors have a responsibility to perform due diligence and it can be very embarrassing for executive management if no preliminary work on block layouts has been accomplished prior to the presentation.

In addition to the time line overlap of planning phases, some generalities do exist in industrial space planning. The following is a sampling of some of the more common ones.

- A building shape that is closer to being square is more flexible and functional (also less expensive) for light manufacturing than a building that has a high aspect ratio of length to width. However, this may not be the case if overhead cranes are required.
- If at all possible, avoid placing a crane bay perpendicular to the direction of facility expansion (see Figure 2-3).
- If your building plan is constrained and you must save space, consider placing some supporting activities on a second floor. Typical single-level building construction costs will rise only 20 to 28% for the same footprint with an additional 12 ft (3.66 m) of headroom (assuming good soil/foundation conditions).
- If pallet-racked storage will be required, plan a building height that allows at least five-pallet-high storage with 36 in. (91.4 cm) of clearance between the fire sprinkler heads and the top of the highest storage point. This will allow full use of standard, narrow-aisle-reach trucks and save a considerable amount of space when compared to using counterbalanced fork trucks. Also consider a second-floor flexible manufacturing level with a "feed" to a first-floor assembly area if floor loading limits permit.
- Generally, storage levels above five pallets high will require what are commonly called "super flat" floors which are more expensive than standard slab floors.

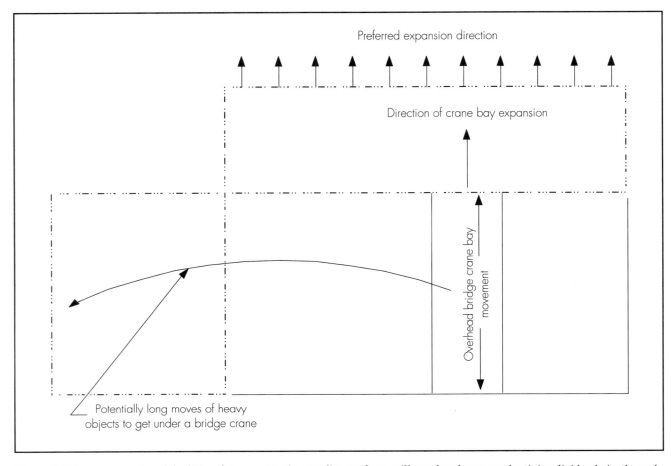

Preferred expansion direction

Direction of crane bay expansion

Overhead bridge crane bay movement

Potentially long moves of heavy objects to get under a bridge crane

Figure 2-3. *In crane-equipped facilities, future expansion studies up front will pay handsome productivity dividends in the end.*

- Add at least 30% to whatever length you have *ideally* planned for an assembly or automated processing line (APL). Use the additional space for an easily-moved function. Never unduly constrain the length of an APL, if at all possible. It is a fact that assembly lines will almost always grow with added operations that must be "squeezed in." Also allow enough space for a buffer loop after critical operations (see Figure 2-4).
- When placing equipment in a floor plan, *always* show all of the maintenance access doors open with all internal shafts or other large machine components pulled out to their full length for removal. This will help avoid a situation where an entire machine must be moved just to service another machine.

- If possible, use vertical power columns (Figure 2-5) to perform the double duty of vertical transportation and buffer storage.
- Shipping and receiving dock doors should not be aligned with prevailing winds, particularly in northern climates. Also, if planning a new facility, try to locate the docks on a side of the building that is not blocking the direction of expansion. Although docks can be moved for an expansion, the added costs and logistics problems of such a change can be avoided with proper planning.
- If you will be expanding an existing building whose floor slab is at or near grade level, and need additional 4-ft- (1.22-m-) high docks for over-the-road trailers, make sure you review the "as-built" construction drawings. It is possible

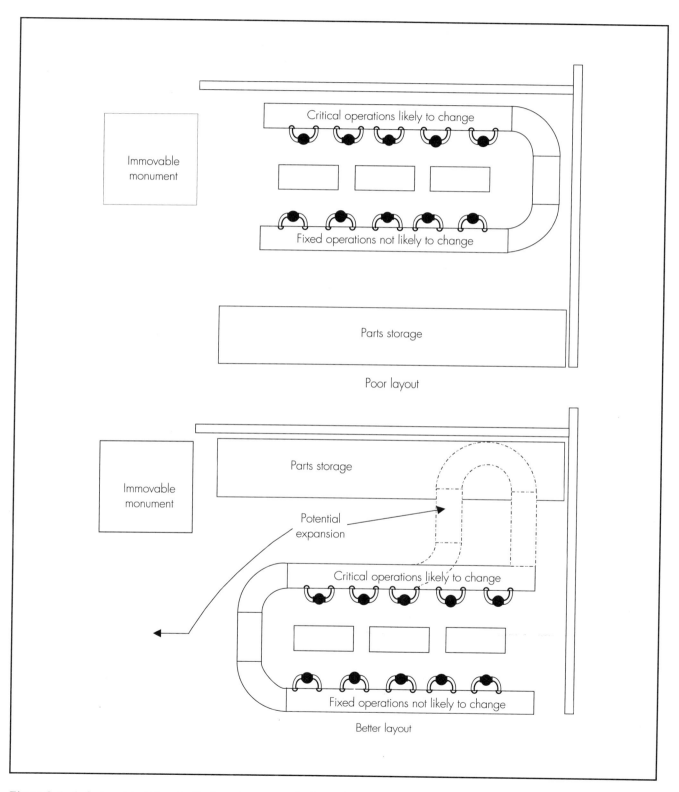

Figure 2-4. *A designed-in future buffer loop serves as a "safety valve" for potential expansion.*

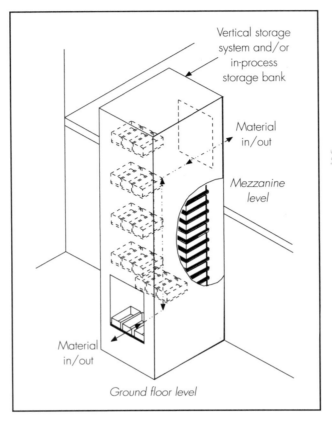

Figure 2-5. Vertical power columns can serve not only as a transport system, but as a temporary buffer storage location as well.

that the required excavation may cause an expensive (and dangerous) loss of floor slab support if the excavation extends below the existing foundation/footings (see Figure 2-6).

- If the plant will be operating on a JIT material receipt basis, you will save some work-in-process (WIP) storage space but you will also need to *increase* the amount of space that is transaction-dependent, e.g., incoming truck unloading and staging areas.
- One of the most important early decisions that has to be made regarding activities and space is whether or not the receiving and shipping dock functions should be combined or separated. Most large companies that have a great number of daily transactions (and can justify full-time crews for each activity) separate them. (In some industries, such as pharmaceuticals, federal regulations or good manu-

facturing practices require the activities to be separated.) On the other hand, most small companies tend to combine these activities so that they may share personnel, equipment, and dock space. This decision will have an extremely important effect on the space layout, since shared docks will result in an overall U-shaped flow within the plant.

- In a new building design, always try to get as much clear span construction as possible to provide maximum flexibility. If the building is located in an area with relatively low snow loads and low seismic activity, per-square-foot cost penalties for a 115-ft (35-m) clear span can be less than 2% (when compared to using building columns on 30-ft [9-m] centers).
- In a new building design, select an area for the mechanical rooms (chillers, air compressors, boilers, etc.) that will not compromise material handling flowpaths and major aisle patterns. If at all possible, avoid locating mechanical rooms in the center of the plant; they should be placed where they do not block future expansion.
- Typically, main aisle space in a light manufacturing facility accounts for between 10 and 18% of underroof space (see Figure 2-7).
- Always plan straight-running aisles, *without turns*, if at all possible. Every time a piece of mobile material handling equipment must make a turn, there is added cost.

This overview was by design presented before the sections on site selection and evaluation of alternative sites. The intent is to highlight at least a few of the factors that need to be considered before the architectural and site-selection process. The overview at this point will help us avoid the potential layout errors cited in this section. We discuss space planning in more detail in a later section.

SITE SELECTION

Establishing a list of potential site candidates usually can be done in a relatively short period of time. To avoid being deluged with telephone calls from industrial real estate brokers and to maintain strict confidentiality, it is best to work through a reputable consulting firm. The consultant will represent you in the "blind," so to

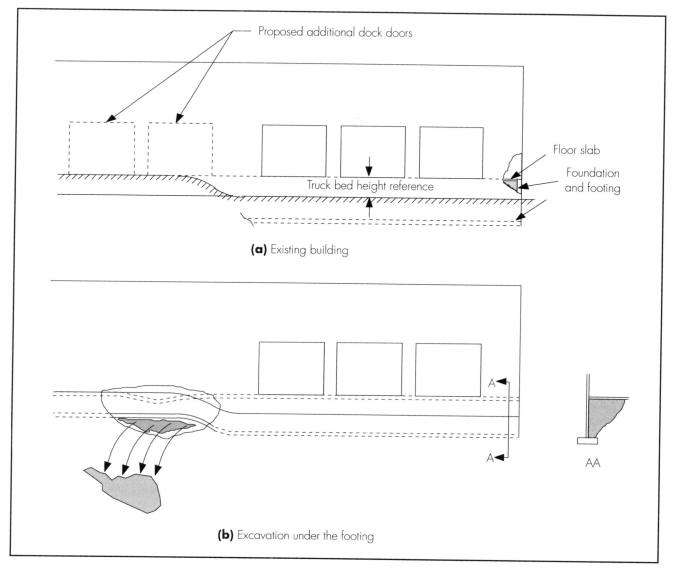

Figure 2-6. *An integral part of adding loading docks to an existing building is the task of checking the as-built drawings to ensure the proper depth foundation and footings are in place.*

speak, and never divulge your company's name or other identifying attributes without your permission. You will, of course, have to pay a professional fee, but the strict confidentiality and avoidance of annoying state and local officials' visits is usually well worth the expense. If confidentiality is not an issue, the best sources of "free" information on available sites and industrial buildings are electrical power companies and rail companies. Of course, they will only tell you about the sites they service, but they are nonetheless still very helpful. The next best helping hands are the area chambers of commerce and the state government or provincial/regional development offices.

Before approaching service organizations or government officials you should be armed with some knowledge of state tax differentials and differential utility rates. I have worked with some manufacturers (large injection molders, for

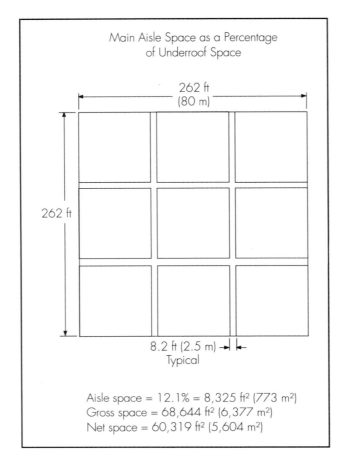

Main Aisle Space as a Percentage of Underroof Space

262 ft (80 m)

262 ft

8.2 ft (2.5 m) Typical

Aisle space = 12.1% = 8,325 ft² (773 m²)
Gross space = 68,644 ft² (6,377 m²)
Net space = 60,319 ft² (5,604 m²)

Figure 2-7. Major aisle space and configuration concerns are always a major issue in facility planning.

example) whose electrical power costs were a larger portion of total product cost than labor. Often, finding a suitable site may involve searches of several regions in a number of states. It is best to narrow your "wish list" to a certain area or region of a state before contacting professional or service organizations for help. When you do, you should at least be able to answer the following site questions.

- Approximately how large a site and building do you need? (Also, most companies request a site with level to gently sloping terrain—not in a 500-year flood plain.) It is not unusual for a new factory site to be six or more times larger than the expected underroof space. Parking areas for light manufacturing alone may consume an area equal to 40 or 50% of underroof plant space.

- How close to an interstate highway access does the site need to be? Can the site be easily serviced by competitive freight carriers?
- Does the site require rail-car access? If so, do you have a preferred core line provider?
- Do you require access to a waterway (rivers or harbors)?
- How close to an airport or express carrier hub does the site need to be? What is the latest airport cutoff delivery time for overnight shipments?
- Approximately how many skilled and unskilled workers will be needed?
- Are factors such as labor unions and state *right-to-work laws* to be considered?
- What are your sewer requirements?
- What is the expected outflow (average and peak) and what is the chemical makeup of your raw process wastewater?
- How many gallons per day of water will your site require and what is the peak demand?
- What will be the electricity demand (kW) and voltage? What is the expected monthly usage (kWh)?
- Approximately how many cubic feet or therms of natural gas are required per day?
- How much solid waste will be generated per day and what is the makeup of the waste?
- Are there any special site requirements? For example, some high-tech companies and precision instrument companies require a vibration-free environment. Other companies may require a special research or high-tech image and would not accept a site near "smokestack" factories. Still others may have a requirement to be within an hour's driving distance from a major airport or university. Also, many companies today need to be located within a defined distance of a major supplier or customer.

After receiving available site information, the next step is to narrow the search to a short list of potential candidates. The next set of questions are more specifically oriented to the short-list sites.

EVALUATING ALTERNATIVE SITES

After you have developed your short list of candidates, you will need to cull those to a smaller list—usually three or four prospects. For each of the remaining prospects, you should obtain the following information for evaluation.

- What is the available labor pool within 15, 25, and 40 miles (24, 40, 64 km) and what *percent change* has taken place over the last 10 years?
- What are the prevailing labor rates for the types of skill levels required?
- If other manufacturers are within the general area, try to elicit their absenteeism rates and employee turnover rates. (Frequently, this information is available through local or regional industrial management associations.)
- Has there been a record of union militancy or industrial strikes in the area? You may want to poll local businessmen and executives to determine the history of the area's labor/management relations and prevailing labor/management attitudes.
- Are governmental training assistance programs available?
- What governmental incentives, tax abatements, and financing plans are available? (Much of this will not be made known until the company is identified and the government officials are convinced that a decision is near.)
- What is the history of the site with regard to past environmental practices?
- What are the specifics regarding short- and long-term capacity availability and reliability for water, sewer, gas, electricity, and waste disposal? What is the area's power outage history?
- What are the total acquisition and site preparation cost differentials for the alternative sites?
- What are the total taxing differences among alternative sites?
- What are the total operating cost differentials among the alternative sites? (Compare labor costs with fringes, truck transportation costs to customers and from suppliers, annual utilities costs, and taxes. Also compare one-time costs associated with employee relocations.)
- What are the attitudes and values of the local community and how do the local people feel the new factory will affect them?
- What are the quality of life differentials between the alternative communities? How do the area costs of living compare (housing costs, schools, etc.)?
- Are local trucking services available?
- Are ambulance services available? How close is the nearest hospital or emergency medical treatment center?
- Are good public schools available?

As can be observed from these questions, there are numerous quantitative and qualitative factors involved with site selection. There are community and local service-related issues that we have not included in this brief overview. However, selecting a site that will benefit your organization for many years into the future takes much more than a comparison of quantitative data; it takes *vision*. This is particularly true if there are no other manufacturing facilities in the area. You can never totally rely on zoning regulations because they can be changed at a later date by newly elected officials. The person or persons doing the evaluation of qualitative factors needs to think as far into the future as possible.

As just one example, a storm water retention pond will probably be required for most manufacturing facilities in the U.S. Some sites offer natural depressions that can help facilitate this need. Other sites may require innovative (and costly) ponds to be constructed (which may also serve to beautify the site or building entrance area).

Environmental assessments also play a large role in site selection. Storm water discharges and site drainage are important factors to be considered. Any water used directly in the manufacturing process, air conditioner condensate, noncontact cooling water, and vehicle or equipment wash water must be captured and treated. In the U.S., it is illegal to discharge these waters or any sanitary waste to a storm water collection system without a National Pollutant Discharge Elimination System (NPDES) permit. The only nonstorm water discharges allowed are:

- Those from fire-fighting activities and hydrant flushing.
- Potable water sources and waterline flushings.
- Irrigation drainage and discharges from springs.
- Lawn watering.
- Uncontaminated groundwater and uncontaminated water from foundation and footing drains.

• Routine exterior building and pavement washdowns which do not use detergents or other compounds and where spills or leaks of toxic or hazardous materials have not occurred.

Environmental factors aside, a very important factor to consider is the overall topography of the site. One of the most difficult pre-purchase cost estimates (without performing a detailed civil engineering analysis) is the cost of excavation and grading work. The amount of earth "cut and fill," as it is commonly referred to, can be very costly. I have seen the costs of site work (earth work, and site prep, paving, utilities, and landscaping) on poorly sloped sites equal 40% of total construction costs! Clearly, unless there is some process-oriented reason for using a heavily-sloped site, a near-level site is preferred, with one notable exception.

Since normal shipping and receiving dock heights for over-the-road trailers are approximately 48 to 52 in. (122 to 132 cm) above the pavement (Figure 2-8), some companies prefer a slightly sloped site to take advantage of a drop in grade level towards the docks. This provides several advantages compared to buildings with

depressed or recessed truck pads. Depressed pads are sometimes difficult to drain and keep clean, particularly in northern climates where freezing occurs. With the proper design, a dock which is not depressed avoids the situation shown in Figure 2-6. The sloped-site strategy requires much more fill and foundation work and is usually more costly; for large facilities, it may be prohibitively expensive. But large projects usually have enough space available to mix both strategies—the plant slab mostly near grade level coupled with a very large contiguous depressed and paved area for trucks (be careful to avoid the condition of Figure 2-6). Also recognize that most of these sloped site strategies generally cause shipping and receiving docks to be located on one side of the building. This in turn results in a U- or C-shaped overall material flow pattern within the plant.

Many other questions and comparisons can be involved with site selection than can be detailed in this overview. Obviously, your investigations also should include a review of plat drawings, soils, and topography, as well as physical site and community walkabouts—each site should be walked by the reviewers at least twice.

OVERALL SITE PLANNING

Overall site planning should overlap with the block layout planning phase as shown in Figure 2-1. Classic space balance and site-saturation methods also should be employed to show the relative gross blocks of space within the facility *and* their relationship to one another. An overall gross material flow pattern should be developed. Also, an allowance should be made on the site drawing, at least in dotted-line fashion, for the direction of *the next* expansion (the one *after* the current expansion).

If site conditions allow, consider the establishment of an inbound corridor or several corridors for primary utilities distribution.

At this time you should also review the easements, green-space requirements, and zoning setbacks associated with the site. Hopefully, you have not chosen a site that lies on a 500-year flood plain, but if you have, the boundaries of the flood area need to be clearly defined.

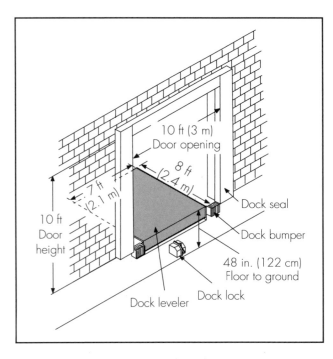

Figure 2-8. A typical truck dock configuration.

One of the main concerns at this stage is automobile and truck access and egress to the site (and rail car, if required). Local fire codes probably require an access road for fire trucks completely around the proposed structure (a gravel road is usually acceptable).

For safety reasons, employee automobile traffic should be separated from truck traffic, if at all possible. Likewise, truck and auto traffic should not cross employee walkways.

On sites located within the U.S., when showing truck direction and turn-in radii on your drawings, be sure to show the trucks entering the area from right to left, in a counterclockwise forward flowpath, and show the trucks backing into the dock in a clockwise, reverse flowpath. This allows the truck driver to directly view where he or she is going in a backward direction. If you lay out your flowpaths in the opposite directions, the drivers are not able to see directly where they are backing and have to rely on viewing their reverse path on their view from the right-hand mirror. This is less safe than the first pattern. On sites located within the U.K., Japan, and other countries where the driver is on the right-hand side of his or her vehicle "flip flop" or reverse the U.S. directions.

On U.S sites, paved truck aprons should extend at least a continuous 120 ft (37 m) outward from the face of the truck docks (135 ft [41 m] is preferred). If the "wrong" direction of truck flow is used, the maneuvering area must be increased significantly, typically adding an additional 70 ft (21 m) to the apron. Some companies prefer to have shorter trucks dock in a separate area to save on apron costs. To be most flexible, however, it is wise to plan the truck road locations, turning radii, and aprons to suit the longest tractor-trailer rigs projected to be driven on the site.

Within the boundaries of a U.S. manufacturing site, plan on a minimum one-way/one-direction truck road width of 13 ft (4 m). Plan a minimum road width of 26 ft (8 m) for two-way truck traffic.

Another major factor to consider in overall site planning is parking. Lack of proper planning for parking areas is one of the *most common* errors in site planning. For a *very gross* preliminary estimate for a manufacturing site in the U.S., multiply the expected total number of shift employees at the plant start-up by 285 to get the approximate number of square feet to allow for a parking lot. If you expect the number of employees to grow by, say, about 8% per year, use 385 as the multiplier to determine the approximate parking space needed 5 years after plant start-up. (In the U.K. or continental Europe, these numbers need to be adjusted downward to account for smaller automobiles and narrower access and egress roads.) When you develop more detailed site and facilities plans, these numbers should be changed to reflect the actual parking "stall" layout patterns. Again, use these numbers only as a starting point to help get the gross site plan started. Also, keep in mind that, in some communities, space allowances for parking are regulated by building codes.

BLOCK LAYOUT PLANNING

Establishing an efficient and effective block layout for a manufacturing plant is a complex task because you are dealing with a complex production process that is constantly undergoing improvement and change. You are really trying to take a snapshot in time, at least 3 years into the future, of a constantly moving target. Sometimes it is difficult enough to take a snapshot of today's operations, let alone take an educated guess at what those operations may look like in 3 years. In some high-tech industries, products and processes change so rapidly, it is even more difficult to make projections — but not impossible.

Normally, if you are planning a major rearrangement of an existing plant, you will be developing a plan hampered by many physical and nonphysical constraints. It is usually neither practical nor economically feasible to try to correct all of the blunders and sins of the past with one plant rearrangement. You need to prepare yourself psychologically for making the best of a less-than-ideal situation. As an example of a typical constraint, most manufacturing plants have "monuments" to take into consideration.

Monuments are those functions, equipment installations, or physical building features that upper management has directed *you will not change* and must find a way to live with. There can be several reasons for having monuments, but usually they can be traced back to economics, risk, and government regulations. Typical

#7

monuments might be the very expensive paint system that was installed last year; the drop hammers with special foundations and pits; the computer test installation that only one person can troubleshoot and repair; the die casting department; the automatic storage and retrieval system that cost millions of dollars to install; the mechanical room (boiler, chillers, compressors, etc.); the solvent recovery department; the new office block; the 3,000-ton punch press and pit; the heat treating department; the micro-etching lab with the extremely expensive vibration-isolated floor; the chemical plating tanks; the enamel or glass frit bake oven; the centralized and newly-tiled and refinished employee restrooms; the freight elevator(s); the shipping or receiving docks; the trenched chip recovery system; the area the company's chairman has reserved for unknown and as yet unnamed "new" products; the company founder's original workshop; and the road or creek that bisects the site. As ludicrous as some of these monument examples sound, all of them have been taken from real-world facilities expansion projects.

On the positive side, there is at least one advantage to rearranging an existing plant—at least you know many of the physical constraints that will affect the plan. That is not always true when trying to develop layouts for a new plant.

When working with block layouts for new plants, you are really working between two unknowns. As demonstrated previously and repeated in Figure 2-9, the block layout-design phase time line falls between site design and detail layout design. At the beginning of the block layout design work, you know quite a bit about the potential site needs, but the actual site has not yet been selected. The site alternatives and constraints will undoubtedly change dur-

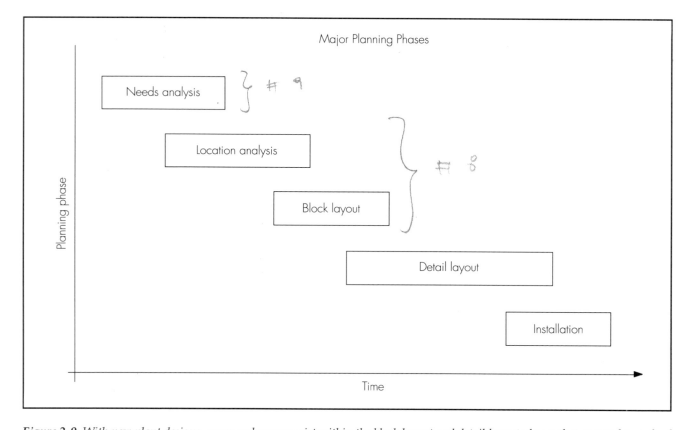

Figure 2-9. *With new plant designs, many unknowns exist within the block layout and detail layout phases that cannot be resolved before a site location is determined.*

ing the course of site selection. At the same time, you know something, but often not a great deal, about the detail equipment requirements, manufacturing cell, and departmental needs. You do not usually know enough to start a detail layout, so the block layout design phase cannot be finalized until almost the end of the time frame allotted for this block planning phase.

During the development of block layouts, exact pickup and setdown points of materials are usually not known. Neither are all of the area shapes or aspect ratios of the areas (the length of an activity area divided by its width). Aspect ratios frequently depend on the configuration of special equipment, length of an assembly line, etc. In establishing relationships among activities, you will probably be forced to use "centroidal" points for each area. Be careful, however, about using centroidal area points as a *total* basis for departmental, cell, or block layouts, because material flow seldom originates and ends at the centroid of a particular functional area. The facilities planner may be lulled into thinking that he or she has a relatively good layout, when in fact it may be a poor layout when the detail layout phase commences and the actual material flowpaths are charted (see Figure 2-10). Again, this reinforces the need to have overlapping phases for block layout and detail layout. We come back to this point again when we discuss detail layout planning.

One of the other major facets of block layout planning (the one that causes its share of grief for layout planners) is developing the "first-pass" area allowance for main aisles. Two basic methods are used for this. (Keep in mind that this first pass is only an estimate that will need to be adjusted later.)

1. After summing all of the individual activity area requirements to get a net total space, calculate a percentage of the total and add this amount of space as a separate item for main aisles, or

2. Take each individual net activity area and multiply each by a major aisle factor (e.g., 1.12) to get a gross activity area space with the built-in allowance for main aisles.

These calculations will give you an allowance for main aisles only and *not the aisles within* each functional area (see Figure 2-7). Typical *net* space

multipliers for main aisles in manufacturing areas run between 1.11 and 1.22.

When calculated against gross areas (where the aisle space is included in the total denominator), between 10 and 18% of gross underroof space is devoted to main aisles in typical manufacturing plants. For example:

Total *net* underroof activity space = 100,000 ft² (9,290 m²)

Multiply by 1.11 to get a main aisle allowance of 11,000 ft² (1,022 m²)

Total underroof space, therefore = 111,000 ft² (10,312 m²)

The same amount taken as a percentage of gross underroof space yields:

11,000/111,000 = 10% (rounded)

If the products or parts being produced are larger than a cube of 4 ft (1.22 m) on each side, the larger end of the range should be used as a starting estimate. If the products or parts being produced are very large, generalized factors do not hold and aisle space can be much larger than the ranges shown.

Most planners prefer to use the method noted in (1) above. By using this method, the layout planner does not lose sight of the individual net activity area requirements, which eliminates potential distortion of individual activity areas. Although the method noted in (2) is easier to use from an overall planning point of view, it can lead to area distortions, particularly with deep or long departments where only the short dimension of the activity block abuts a main aisle (see Figure 2-11). The method in (2) also tends to favor those long and deep activities with more space than their actual needs dictate and, as a result, tends to produce layouts that are short of aisle space elsewhere.

Flexibility must be considered where the block layout phase ties into the total building plan and site plan. It is absolutely essential that the planner have at least some industrial engineering or manufacturing engineering knowledge of the company's manufacturing processes. This does not mean that the layout planner needs to be skilled in all of the processes, but it does mean that he or she must have some knowledge of where the "choke" points or bottleneck processes

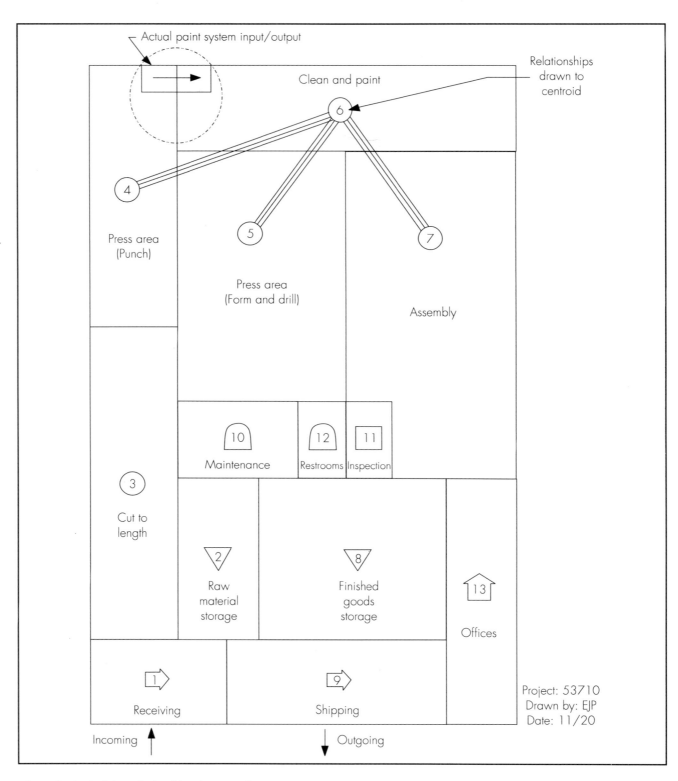

Figure 2-10. *Activity relationships drawn to "centroidal" points can result in poor overall layouts. The detail layout, however, will determine actual material flowpaths.*

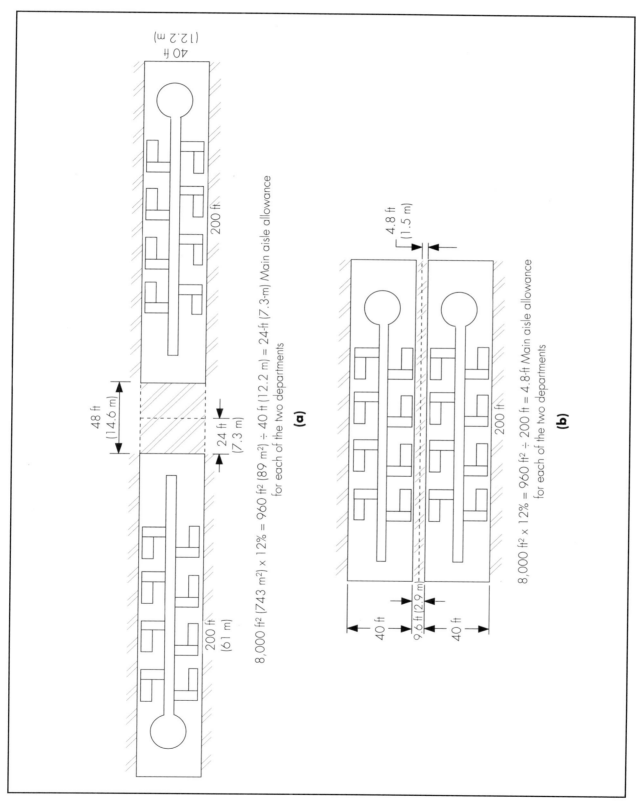

Figure 2-11. *Depending on the method of calculation, area allowances for main aisles can result in a poor layout (a) or a functional one (b) with no area distortion.*

are located. One can have a nearly perfect layout based on material flows and other relationships between activities, only to find out later that the choke activities have been boxed in or located along a side of the building that cannot be expanded. A "sensitivity for expansion" analysis (described later) needs to be performed, preferably before we begin to develop block layouts.

SCHEMATIC UTILITIES/ENERGY PLANNING

For companies with existing operations, the most useful information for utilities and energy planning is a database of actual historical usage, structured in *energy unit usage* and not costs. Costs can be used secondarily for reference ratios and projections, but costs cannot help the planner determine detailed service requirements in any detailed fashion.

The planner needs to establish a database of existing "nameplate" equipment or equipment specification requirements. Each piece of electrical equipment, for example, should have a nameplate indicating power, amperage, voltage, hertz, etc. For a new plant, supplier specifications for the proposed equipment need to be used.

Usually, there are a significant number of electric motors, air conditioning systems, air compressors, etc., in the plant, so you also need to obtain information on the historical or expected power demand factor. These inductive electrical loads cause (what appears to be) a power surge or a surge in capacity for which the power companies must size their incoming service lines. Power companies record the highest observed peak power used, which is usually much higher than the sum of the actual resistive kilowatts (amperes multiplied by voltage) pulled by the connected loads. Technically, this is due to a phase difference between the instantaneous voltage and current levels caused by the inductive motor loads. This peak typically occurs on a hot summer day when many machine motors are started at the same time and air conditioners are running at maximum output. A factor is developed for this observed peak and all of the user's actual power usage for the next 12 months or so is multiplied by this factor, even though that

peak may only have been observed for a split-second or so. In essence, most companies pay a penalty all year for that split-second peak. As a result of these penalty costs, which can be very high in some industries, many companies institute staggered equipment start-ups and install devices such as capacitor banks (which almost exactly counter the effects of inductive loads) to lower the observed peak and bring it closer to the actual resistive kilowatts used.

Compressed air is another secondary energy item to consider. I remember doing a master site and plant layout project for a U.S.-based drop forging company which employed six 100-horsepower air compressors located throughout the plant. They were operating continuously at full capacity. The company also employed several stand-by diesel-powered compressors in the event one of the main compressors was shut down for service. The plant operated 24 hours per day. Company management knew it had some problems with high energy costs in general, but had no idea what energy losses they were incurring due to poor piping design, leaks, etc. After several years of continuous production, the company shut the plant down completely for one weekend to do some major repairs unrelated to the compressors. With the plant shut completely down, everyone expected an eerie silence instead of the banging of the drop hammers. Not so. With *all* of the production equipment shut down, one of the main compressors was still running continuously at full capacity and another compressor was cycling on approximately 15 to 20% of the time *just to keep up with the leaks in the piping system!*

A similar company in India corrected air losses, saving the company 10% of its energy costs for compressed air. Almost half a page of the Indian company's annual report was devoted to this savings in compressed air energy costs.

The point is that common delivery systems such as air compressors need to be centralized in relation to the air usage pattern. Pipe runs should be as straight as possible and as short as possible with the fewest number of connections.

Similarly, in an effort to hold down installation costs and avoid unnecessary pressure or voltage losses, plan for efficiency. Piping and conduit systems for *all* utilities whether for electric, water, gas, steam, storm water, sanitary sewage, or effluents such as industrial waste and exhaust fumes should be short and straight-running, with as few connections as possible, (notwithstanding any necessary traps).

When using a computer-aided design/drafting (CAD) system, it is customary to lay out the various utility plans on separate drawing layers. For example, there would be one CAD overlay for the electrical installation, one for plumbing, one for the fire protection system, one for sewage lines, etc. In addition to the separate utility layers, each one of which shows the appropriate above- and below-ground installations, there should be at least three combined overall site overlays. One to show all of the above and below-ground primary utility feeds to the building(s) on the site, another to show the combined underground secondary utility feeds under the floor, and a third overlay to show the combined above-ground secondary utility feeds above the floor. Although software is available that helps prevent utility run interference, these combined overlays are very important when designing materials handling (e.g., overhead conveyor systems and high-bay storage) systems.

If you are working on the rearrangement of an existing layout and need to cut into the floor or ceiling, it is extremely important to get certified "as-built" drawings from whomever was responsible for the construction of the building. There have been more than a few instances where workmen were electrocuted when sawing through concrete floor slabs and cutting into buried power lines. The building design drawings showed the power lines in the floor to be some 6 or more feet (1.83 m) away from where the work was being done. However, the power lines had been shifted during the construction process (for some unknown but probably valid reason) and the design drawings were never brought up to date.

As in the detail layout phase, it is very important to keep all supervisors and other interested parties informed of the block layout process. After block layout approvals, significant changes should receive signoffs from upper-level managers.

DETAIL LAYOUT PLANNING

To many novice layout planners, developing the detail layout plan is the beginning of the process. Owing to a lack of knowledge, they tend to skip entirely the site and block planning phases. Even some experienced layout planners who have never been exposed to top-down planning fall into this trap. The bottom-up detail layout approach may work for small projects but can prove to be an overwhelming nightmare on large projects. In my experience, the most disorganized plant layouts can usually be traced back to starting from the bottom and working up.

Problems occur frequently when you have several separate focused teams working on small detail portions of the project and no experienced individual in charge of coordinating the project from the top down (i.e., responsible for site and block layouts). In most instances, the use of "self-managed work teams" without the benefit of an experienced overall project manager has proved to be a mini-disaster when planning large layout projects. The total overall picture is easily lost as each team strives to optimize its own little corner of the world, sometimes to the detriment of the entire enterprise. After doing several large projects, one learns quickly that it is far better to work from the top- or gross-level planning stages to the bottom detail rather than the other away around. The most successful layout planners understand that the site and block layouts should be started before detail layouts. Some planners would argue that the block and detail layouts should be done concurrently if resources allow.

The most obvious and common errors associated with doing detail layouts first (before block layouts) is the lack of an overall materials flow and logistics plan for the plant and site (e.g., location of receiving, shipping, and materials staging areas). Differently-focused teams, unless managed well by an experienced coordinator, invariably develop suboptimum logistics plans *for the plant as a whole,* even though the layouts for their individual respective areas may be very good. Another problem that surfaces when working from the bottom up is the lack of thought given to common personnel and plant services such as lunch/break rooms, maintenance areas, restrooms, water fountains, etc.

That warning aside, the development of detail layout plans will usually be accompanied by some level of methods improvement if you are rearranging an existing plant. We touch on this later in the book when we discuss operations process charts and manufacturing cell design examples.

Detail layout planning follows the same general pattern of procedures as block layout planning. The level of detail information, particularly on physical plant and equipment specifications, is at a fairly high level.

The detail layout process requires considerably more time and therefore generally costs more than the block layout phase. Although detail planning requires a relatively high level of mechanical aptitude and visual skills, it is usually delegated to members of the organization who are at a lower level in the organization than the individuals doing the site and block layouts. The main reason for this is that a much closer interface must exist between top management and the individuals planning the site and block layouts. This interface is normally not required at the detail level unless a very sophisticated or entirely new manufacturing system (new process cell, new assembly line, etc.) is being installed and upper management wants to keep tabs on its progress.

Just as there must be good communication and coordination with upper management on site and block planning, detail layouts require a heavy emphasis on communication and coordination among individual area and function supervision, production personnel, and the detail layout planner.

The mechanical process or logic used in detail layout planning within specific departments or work cells is very similar to the process we use for developing block layouts. Instead of relationships being developed between blocked-out areas, they are now developed between individual operations and/or machines. In the detail layout, the location of a time clock, or *kanban* visual board, or cabinet of supplies may be included in developing relationships. The detail layout of a production area should show each machine (with maintenance access doors swung open), each operator, materials setdown area, materials pickup area, auxiliary equipment, benches, aisles, etc.

Operational process charts or routing sheets are almost a necessity for understanding intradepartmental material flows. Each worker's space, equipment, and materials setdown area have to be accounted for. Library parts pre-programmed in your CAD system for individual pieces of equipment or equipment footprint templates (if a manual layout method is being used) become quite useful at this point.

The detail layout efforts may be enhanced significantly with the use of dynamic computer-based simulation tools. Experienced layout planners often use a dynamic simulation program as their primary layout tool. Expected throughput and production rates can be simulated for various detail layouts and verified before completing the final design. Space requirements for buffer storage or space needed to handle critical equipment breakdowns can be estimated in advance. However, it is very difficult to attempt to simulate an entire plant's operations. It is much easier to simulate a particular set of operations or a cell within a small sector of the plant. That is where dynamic simulation really proves its worth.

Adjustments to the block layout areas are often required during the detail layout phase. Usually these are made necessary because of unforeseen improvements in the detail layout or slight modifications to the overall work center layout.

Erroneous assumptions can be made in both the detail phase and the block layout phase. Equipment assumptions should be researched thoroughly before getting equipment quotations from vendors. Examples include assuming that conveyors can be hung from the roof or ceiling (which is almost always a *wrong* assumption) or that a bridge crane can be mounted on the existing building columns. Any such assumption *must* be verified by a structural engineer. Generally speaking, if conveyors or cranes were not allowed for in the original building design, the existing structures will not support the loads without structural modifications. This is particularly true for any structures that are located in seismically active areas (see Figure 2-12).

It is very important to keep all supervisors and other interested parties informed of the de-

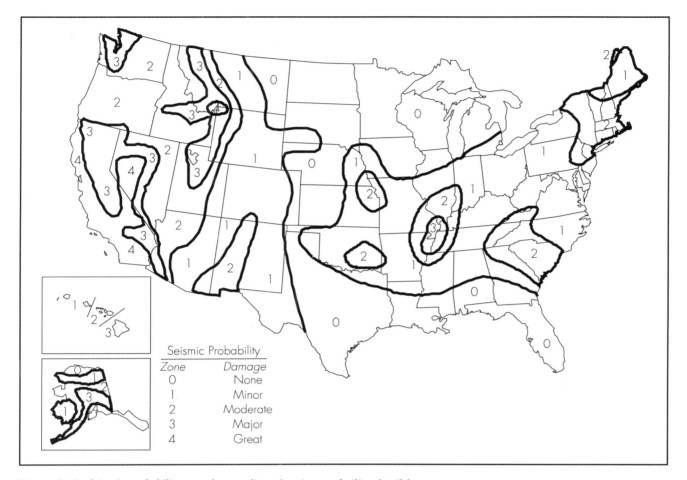

Figure 2-12. Seismic probability may have a direct bearing on facility detail layout.

tail layout process. As in the block layout phase, significant changes should receive signoffs from managers and supervisors.

Approvals for detail layouts should be received from:

- The individual supervisors responsible for each workstation within the area.
- Safety and/or security personnel.
- Supporting service departments.
- The person in overall charge of the operations covered by the layout.

The signoff procedure should come as a result of policy and not an individual's desires. It is important that the procedure be followed so that it does not appear that the planner is trying to put an approver on the spot by signing-off on drawings.

IMPLEMENTATION PLANNING

The key factors in implementation planning are project management and good communications. Successful projects usually have a dedicated, objective champion charged with creating and implementing a detailed project task plan and schedule, and having the authority to carry the project through to completion.

On new plant projects or major rearrangements involving many skilled construction trades, a computer-based project management system (CPMS) with resource leveling capabilities is a definite asset. On very large projects, CPMS is a necessity. Figure 2-13 is a sample schedule for a major rearrangement. For small rearrangement projects, a simple Gantt-type chart (Figure 2-14) may suffice.

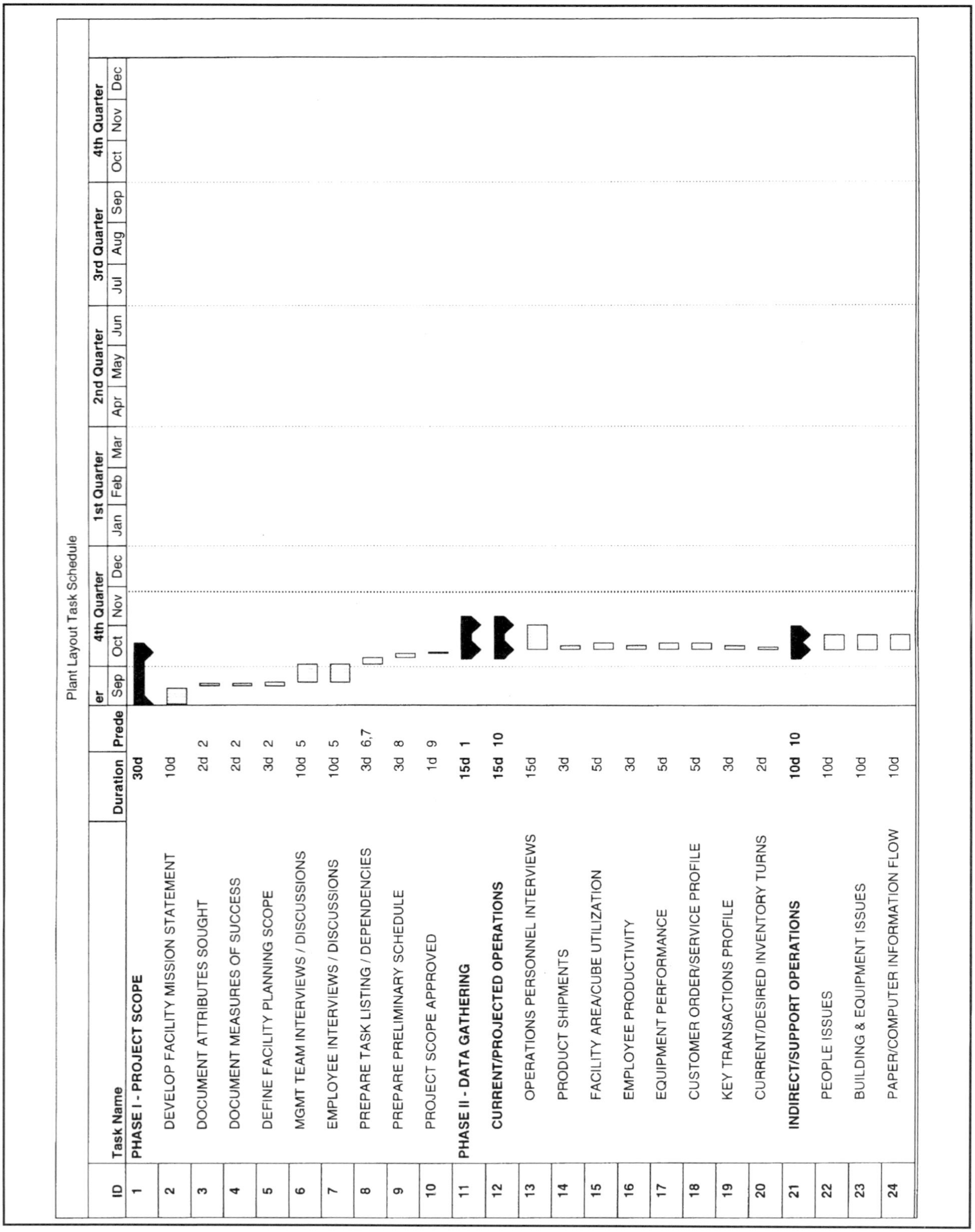
Plant Layout Task Schedule
ID
Task Name
Duration
Prede
4th Quarter
Sep Oct Nov Dec
1st Quarter
Jan Feb Mar
2nd Quarter
Apr May Jun
3rd Quarter
Jul Aug Sep
4th Quarter
Oct Nov Dec
1 PHASE I - PROJECT SCOPE 30d
2 DEVELOP FACILITY MISSION STATEMENT 10d
3 DOCUMENT ATTRIBUTES SOUGHT 2d 2
4 DOCUMENT MEASURES OF SUCCESS 2d 2
5 DEFINE FACILITY PLANNING SCOPE 3d 2
6 MGMT TEAM INTERVIEWS / DISCUSSIONS 10d 5
7 EMPLOYEE INTERVIEWS / DISCUSSIONS 10d 5
8 PREPARE TASK LISTING / DEPENDENCIES 3d 6,7
9 PREPARE PRELIMINARY SCHEDULE 3d 8
10 PROJECT SCOPE APPROVED 1d 9
11 PHASE II - DATA GATHERING 15d 1
12 CURRENT/PROJECTED OPERATIONS 15d 10
13 OPERATIONS PERSONNEL INTERVIEWS 15d
14 PRODUCT SHIPMENTS 3d
15 FACILITY AREA/CUBE UTILIZATION 5d
16 EMPLOYEE PRODUCTIVITY 3d
17 EQUIPMENT PERFORMANCE 5d
18 CUSTOMER ORDER/SERVICE PROFILE 5d
19 KEY TRANSACTIONS PROFILE 3d
20 CURRENT/DESIRED INVENTORY TURNS 2d
21 INDIRECT/SUPPORT OPERATIONS 10d 10
22 PEOPLE ISSUES 10d
23 BUILDING & EQUIPMENT ISSUES 10d
24 PAPER/COMPUTER INFORMATION FLOW 10d

Figure 2-13. For large projects with a multitude of tasks, a computer-based project management system is essential.

Plant Layout Task Schedule

ID	Task Name	Duration	Prede
25	PHASE III - DATA ANALYSIS/DESIGN	40d	11
26	SUMMARIZE/ANALYZE DETAIL DATA	20d	
27	OPERATION PROCESS CHARTS	5d	
28	MATERIALS CLASSIFICATION SUMMARY	2d	
29	MATERIAL FLOW/PUSH VS. PULL/CELLS	20d	
30	QUALITATIVE RELATIONSHIPS	20d	
31	QUANTITATIVE RELATIONSHIPS	20d	
32	SPATIAL RELATIONSHIP DIAGRAMS	20d	
33	DETERMINISTIC/DYNAMIC SIMULATIONS	20d	
34	CONCEPTUAL DESIGN(S)	15d	26
35	GROSS IMPLEMENTATION COSTS	15d	
36	SAVINGS/OPERATING PROJECTIONS	5d	
37	MODIFICATION FACTORS	5d	
38	COSTS & WEIGHTED FACTOR ANALYSIS	10d	
39	GRAPHIC "BEFORE/AFTER" FLOW CHART	10d	
40	SELECT AND REFINE ALTERNATIVES	5d	34
41	PHASE IV - FACILITY REARRANGEMENT DESIGN	80d	25
42	DEVELOP BLOCK LAYOUTS	5d	
43	DEVELOP GENERAL UTILITIES/SITE PLANS	15d	
44	DEVELOP SPACE ASSIGNMENTS	15d	
45	PREPARE PRESENTATION GRAPHICS	5d	44
46	CONDUCT CLIENT DISCUSSIONS	10d	45
47	REFINE PLANS	5d	46
48	OBTAIN CORPORATE APPROVAL	15d	47

Figure 2-13. (Continued)

Plant Layout Task Schedule

ID	Task Name	Duration	Prede
49	ARCHITECTURAL PLANS/UTILITIES	15d	46
50	ENGINEERING LAYOUTS/EQUIPMENT	20d	48
51	EQUIPMENT/CONSTRUCTION ALTERATIONS	5d	50
52	PREPARE DETAILED PHASING PLANS	5d	51
53	PREPARE BID DRAWINGS/SPEC PACKAGES	30d	48
54	PHASE V - PROJECT IMPLEMENTATION	100d	41
55	VENDOR CANDIDATE SELECTION	2d	
56	VENDOR BID EVALUATIONS	15d	
57	CONSTRUCTION MANAGEMENT	80d	
58	CREATE/COORDINATE SCHEDULES	80d	
59	CONTRACTOR MOBILIZATION	10d	56
60	MATERIAL / EQUIPMENT DELIVERIES	40d	59
61	INSTALLATION	15d	60
62	START-UP	15d	61
63	RUN CLIENT/VENDOR MEETINGS	80d	56
64	MONITOR VENDOR PERFORMANCE	70d	59
65	PERSONNEL TRAINING SERVICES	5d	61
66	START-UP ASSISTANCE	15d	65
67	GENERAL DEBUGGING SERVICES	20d	61

Figure 2-13. (Continued)

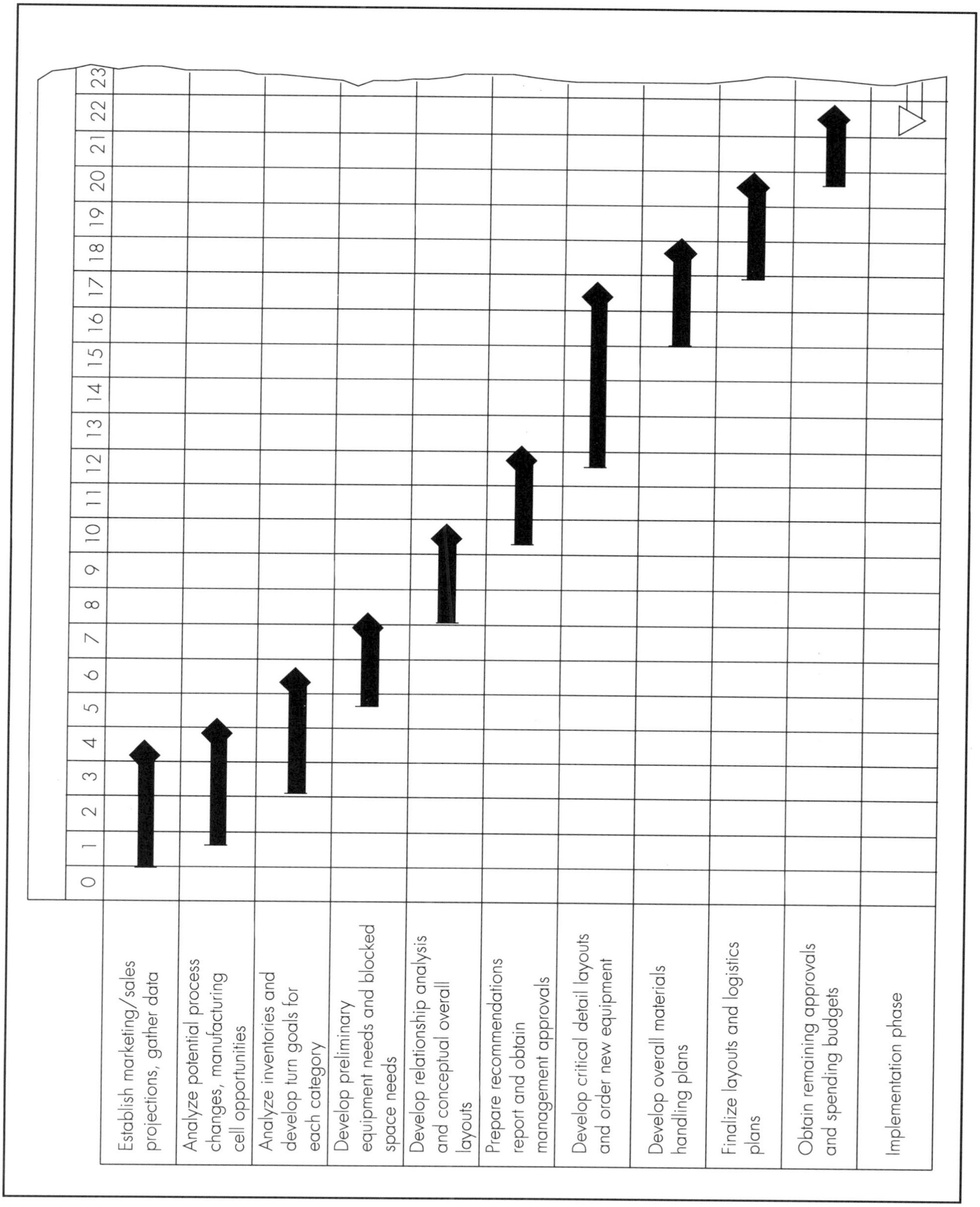

Figure 2-14. For projects that require only a few tasks, simple Gantt-type charts can track progress.

The project manager should have a working knowledge of the interface and sequences required among the various skilled trades and contractors during implementation, including the potential interferences that can occur.

The project manager and members of the project management team are best selected from the original layout planning group. They are most familiar with the reasons and rationale for the layout decisions and the tradeoffs that were made during the analysis. This is a crucial point because major errors can be either introduced or avoided when quick decisions must be made during construction.

For instance, on a relatively small new raw materials warehouse project, a facilities planning consultant was engaged to determine the building configuration requirements. The consultant performed a detailed industrial engineering analysis and growth projection to establish a pallet storage configuration and layout, along with a clear height for the new building. The owner then hired a design/build contractor to construct the building. The contractor informed the owner that he could save him approximately 2% of the total building construction costs (a considerable sum) if the building wall panels and roof could be lowered by 2 ft (0.61 m). The contractor said he looked at the pallet rack specification and stated that "the tops of the uprights would still be below the roof trusses." He said he checked with the city building department and they indicated the building permit would not be affected by the change. The owner jumped at the chance to save some money. Without contacting the consultant who developed the plan, the owner gave the contractor the go-ahead to lower the roof and building panels. It was not until after the building was constructed and the pallet rack was erected that the owner discovered there was not enough building height to store pallet loads on the upper pallet level. Since the consultant had planned to use five-high pallet storage, *the owner lost 20% of his total planned storage capacity.* Had the owner contacted the consultant or kept him involved for answering technical questions, the calamity could easily have been avoided.

Similarly, on a recent large project, a major architectural firm did not bother to coordinate the location of roof drains with the manufacturing engineers who were doing the detail layout of the plant. This was a fast-track project and the detail equipment layouts were being done as the building was being designed (architecturally) and constructed. The roof drains were planned to come down the side of the building columns (I-beams) and be cast into the concrete floor slab (to connect to the drainage system located beneath the slab). The plant layout engineers assumed the drains would be placed as shown in Figure 2-15(a). Major pieces of equipment were to be located within 2 in. (51 mm) of the building columns as shown. You can guess what happened. The drains were placed as shown in Figure 2-15(b). Because the new building was being constructed more than 1,500 miles (2,400 km) away from the existing plant, the company did not believe it was cost effective to have their manufacturing engineers as part of the on-site project management team. After tearing up the floor slab and relocating all of the drain pipes, this company rethought the economics of its travel policies!

In the U.S., one of the frequently overlooked items (and one that should be listed near the top of the task plan) is obtaining environmental permits. Even if "minor fumes" are emitted from a production process you plan to move and the fumes contain volatile organic compounds (VOCs), you need a permit. The permit stipulates the maximum allowable daily or monthly limit of VOCs. The *potential* output for a particular machine is viewed by the government as *the* output even though you may only use a small portion of the machine's capacity. For example, if you have a machine capable of 4-ft (1.22-m)-wide printing or silk screening and you only use a 2-ft (0.61-m)-wide printing area for your products, the VOCs emitted will be calculated assuming you use the total 4-ft-wide area. Let's assume your permit allows you to produce 1,000 pieces per day of a 4-ft^2 (0.37-m^2) printing pattern based on projected VOC emissions for the inks you are using. You actually produce 1,200 pieces per day of a much smaller 2-ft^2 (0.19-m^2) printed product. Based on the surface area of the print, the actual VOCs emitted are only 60% (2,400/4,000) of the allowable. However, even though your machine may *actually* be emitting only 60% of the permitted VOC limit, the government will apply penalties based on exceeding the *theoreti-*

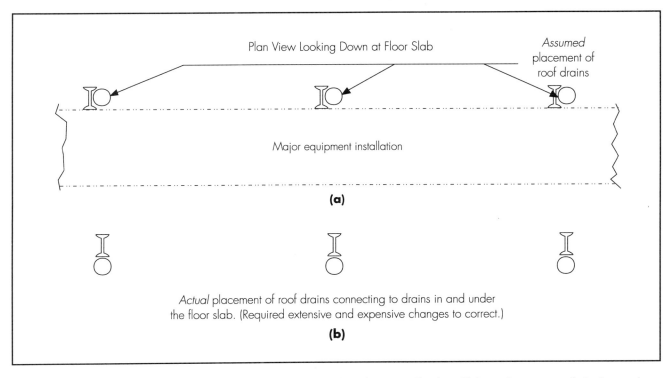

Figure 2-15. Small changes during construction of a facility can have huge ramifications if the project or manufacturing engineer is not kept in the information loop.

cal capacity-based limit. I have seen situations where a printing plant exceeded its total theoretical VOC "bubble limit" and several machines had to be shut down on a regular basis. This shutdown was forced by the regulators, even though the company was not even close to approaching the permitted VOC limits on an actual usage basis.

With older plant renovations, other environmental and building code variances also may become a problem during implementation, even though the affected areas may not be near the area bounded by your plant layout changes or additions. For example, there should be a thorough evaluation of older plants to ensure that no asbestos is present. Local government building inspectors also may require other areas of the plant to be brought up to existing code requirements. As an example, this may require the installation of special toilet facilities, ramps, and elevators for the handicapped as well as other major upgrades. The potential for costly changes to other areas of older plants needs to be thoroughly investigated before committing to a particular renovation strategy.

Not knowing of pre-existing use and soil conditions, some companies shy away from doing any groundbreaking or digging in previously occupied but now vacant areas on their sites. They sometimes are afraid of what environmental problems they may uncover and what unknown liabilities may be involved.

Equipment clearances also are frequently overlooked in existing plant rearrangements or additions and new construction. I have witnessed several situations where new masonry walls were constructed, only to be torn down later when the production equipment arrived and would not fit through the doorway.

There have also been situations where ceiling heights in portions of the new facility were too low. Some companies literally "raise the roof" or dig pits to allow for new equipment. This height requirement for major pieces of equipment is normally known in advance, but on auxiliary equipment, it is sometimes overlooked.

A case in point: the receiving dock area in a new plant was designed for aesthetic purposes to have relatively low head room. A portion of the area was designed to be a lift truck battery charging room. The architect went to great pains to design a well-ventilated masonry- walled room with automatic battery pullers, conveyors, recharge stations, special battery wash showers, and floor drains. The room was designed for the standard counterbalanced fork trucks that he was accustomed to working with on other projects. However, this particular charging room was supposed to handle the company's *storage area lift trucks* as well as the dock area lift trucks. Although he checked the capacities and charging rates of the storage area lift truck batteries, the architect neglected to check the height of the masts on the trucks. These were high-lift, triple-mast fork trucks which, even with the masts fully collapsed, could not fit under the charging room's low ceiling (see Figure 2-16). This one small detail resulted in demolition of the room and a major relayout of the entire receiving, staging, and incoming storage area of the plant.

Despite these horror stories, there are a great many more success stories. The key factor or common thread in most success stories (and absent from the horror stories) is ongoing communication and feedback among everyone involved in the planning, design, and construction phases.

In implementation planning, it is much too easy to focus on the plan view of the floor layout, showing all of the equipment neatly in place. The planner must be careful not to lose sight of equipment clear heights and how, physically, each piece of equipment will be transported, erected, and used in its final position. For large pieces of equipment, this may require that the equipment be in place *before* the walls or ceilings are erected. These conditions should be accounted for in the task sequences of all project management plans.

Timing or move phasing is also a key planning factor. One of the most interesting and difficult facets of implementation planning is scheduling the phased rearrangement or moving of equipment without disrupting production

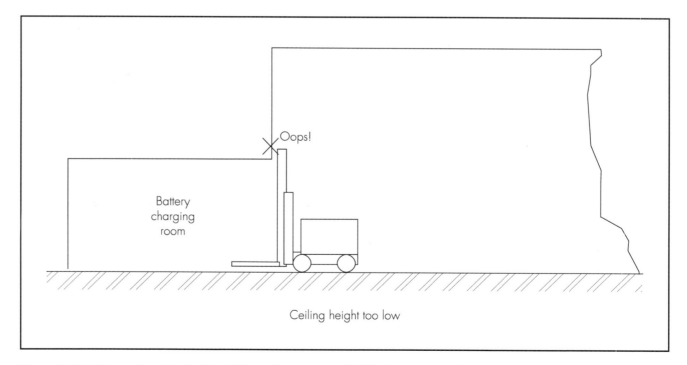

Figure 2-16. A thorough understanding of the equipment a new facility is to accommodate is critical to effective design.

operations. If moving existing equipment, it is generally wise to plan the move of equipment closest to the main doorway or dock first. This will open up a hole for staging of subsequent equipment moves. If that is not possible, at least consider clearing an area in a direct path to the door or opening through which you will be moving the equipment.

Implementation or installation planning is generally a relatively easy task. Usually the actual physical move is a much more complex task. Move details invariably require running changes and compromises. Many on-the-spot quick decisions need to be made during the physical move process. As mentioned earlier, it is preferable to have at least one member of the original layout planning team heavily involved in the implementation. He or she is an invaluable source of information and technical advice and can help prevent wrong and costly decisions.

3

WHERE TO BEGIN

THE NEED FOR CHANGE

New plants and plant expansions can begin with a new product introduction, a great increase in demand for a company's existing products, or an expansion into a new region or even into a different country. Some new plant decisions are responses to a logistically limited site, a gradual change in the quality of the available work force, the inability to recruit a suitable work force for added work shifts, or any of a host of other reasons. However, most expansions or consolidations of existing facilities begin when someone in authority feels an excessive amount of "pain." Pressure may come from a higher level in the company to drive production costs down or to increase production quantities significantly without a corresponding increase in the number of employees. From lower levels in the organization there may be constant complaints about the inefficiencies caused by poor shop-floor layouts or poor facilities. These complaints may be multifaceted. In small and medium-sized companies, it is not unusual for the need for new offices or a newer, more modern corporate image to drive the rationale for a new plant. Obviously, the new plant would include new offices as well.

Whatever the reasons for the proposed changes, a new plant or an existing plant expansion or consolidation offers the company a golden opportunity. That opportunity is centered on the one-time chance of doing things right by implementing improvements to the manufacturing operations and all supporting activities. In some companies, there will not be another chance for 5, 10, or even 20 years to make

major improvements. *Hence this may be the only opportunity the company will have in many years to truly leapfrog its competition.* It is imperative that company management be made fully aware of this opportunity. Companies who seize this advantage to improve operations are usually the leaders in their industries. These leading companies also recognize that there is a definite manufacturing and industrial engineering skill level required for "doing it right."

Executives who do not take this one-time advantage, and who simply ask their plant engineering or plant maintenance departments to duplicate or clone current operations in a new facility are, more often than not, the marginal or poorest performers. Unless they have a proprietary product with very little competition, a decision to clone current operations is usually a mistake. Sometimes, however, a cloning approach is used strategically by industry leaders if they have an aging "cash cow" type of product in one of their divisions and there is no real long-term plan for saving that portion of the business. Sometimes this cloning process is just the result of uneducated or inexperienced decision-makers at upper management levels who may not recognize that there are special skills associated with designing plant expansions.

I recall one case where this occurred with a relatively small Ohio automotive parts division of a larger, well-known, Michigan-based company. The company's customers were two of the Big Three automobile manufacturers. One of these remarked on how poor the supplier's existing plant layout was, that the layout must be contributing to manufacturing inefficiencies, and that he hoped the layout would be improved with the plant expansion. These casual remarks by several customers, led the plant manager to attend a 2-day seminar on manufacturing plant layout sponsored by the Society of Manufacturing Engineers (SME). The process of expanding his plant had begun, and the steel structure was being erected even as he attended the SME program. By the end of the program, he realized that he did not have the time nor the internal technical staff with the experience to properly plan the internal layouts. He had only guessed where the receiving and shipping areas should be and he had not done any analysis on the rest of the internal layouts. He had initially planned to just clone his existing internal operations to increase capacity and output. After the seminar, he asked a consulting firm to prepare a proposal for developing alternative layouts and improvements to the division's materials handling and manufacturing processes.

The plant manager was enthused and, since timing was critical, gave a preliminary go-ahead to the consulting firm to schedule the project. He needed just one signature from the corporate vice president of manufacturing before he could officially give the go-ahead to the consultants. The quoted fee was relatively low and he did not see any problem whatsoever in getting corporate approval. However, when he asked the vice president for the approval, the fee quoted was not even discussed. The plant manager was told, in no uncertain terms, that *if he could not personally lay out the expanded plant, the company did not need him and he would be fired!* Needless to say, the corporate vice president had never been involved with a major plant expansion and had no knowledge whatsoever of the skills required nor the opportunity that presented itself. The plant manager did the best he could but, with all of his other duties, could really not spend the time and effort required to do much more than a cloning operation. A once-in-a-decade golden opportunity for making significant improvements was lost.

On the positive side, there are many success stories. I worked with one U.S.-based company which was building major, custom-designed, replacement portions of a top brand molding machine originally produced in Asia. The U.S.

firm's selling point was top quality and speed of delivery, quoted at 12 weeks for custom-designed machine assemblies. Once started in the shop, and barring any special problems, an order actually took 8 to 9 weeks to fully complete using a traditional batch-process scheduling system. At first the company competed favorably on delivery time, since the original equipment manufacturer was shipping their replacement assemblies by ocean freight to the U.S.; however, the company was starting to feel the pressure when the Asian producer began to stock many standard components in the U.S. and customized them upon receipt of orders. During the planning of the U.S. company's plant layout for a new expansion, engineers discovered that most of the parts produced had enough similarities to warrant investigation of a cellular approach in manufacturing instead of a traditional functional/process-oriented layout. After a thorough analysis, the new layout for the plant expansion was developed into two separate cells. *Processing time through the streamlined shop was reduced from 8 to 9 weeks to 4 days!*

In another success story, an automotive parts manufacturer found that approximately $200,000 per year in materials handling costs could be saved by a proper analysis and placement of one business unit in a plant expansion. This was shown at the block layout stage in materials handling logistics alone, even *before* considering the detail layout. The benefits in time savings were even more important than the cost savings.

With another consumer products manufacturer, changes in the plant layout, coupled with minor materials handling improvements, resulted in *direct labor reductions of 55%, a 225% increase in output per worker, and a 40% reduction in WIP inventories.*

In all cases, the difference between achieving a competitive cost advantage or incurring unnecessary, ongoing, high annual operating costs hinged on *sound business knowledge*. If management does not know (or does not want to recognize) that there are competitive advantages to be gained by taking a professional approach to plant layouts, they will lose a golden opportunity. The applications of analytical skills, manufacturing engineering knowledge, sound judgment, and "hard knocks" experience to develop optimum layouts will pay for the effort many times over.

PEOPLE SKILLS AND LEADERSHIP

Clearly, the most valuable asset one can possess in developing improvements in manufacturing processes and plant layouts is *experience*. There is nothing more educational in this field than making one's own mistakes (and not repeating them). This is particularly true with plant expansions or new plants where major construction is required.

Many organizations today form individual working teams with intimate knowledge of their own particular manufacturing processes. These highly focused teams are then chartered to develop their individual departmental layouts. As discussed earlier, without the benefit of an experienced project manager, this practice can lead to less-than-optimum—and sometimes very costly—results. In my experience, there are 10 major attributes or skills that an *ideal* project manager should have.

1. Prior experience in managing at least one (preferably more) construction or major move project, if a major construction or plant relocation is required.
2. A working shop-floor knowledge of manufacturing processes and their physical constraints, in addition to a working knowledge of overall site and plant logistics requirements.
3. Above average intelligence (an absolute requirement).
4. A reputation for and dedication to excellence and not taking the easy way out of a problem.
5. Broad experience with materials handling equipment and general knowledge of the pros and cons of various equipment schemes.
6. Experience with material flow analyses and developing layouts using activity-based relationships/affinities or other quantitative techniques.
7. An ability to communicate effectively at all levels, from the shop-floor operator and construction laborer to the company's executive level.

8. A keen sense of politics and leadership and an ability to motivate both teams and individuals. Politics aside, and above all else, the project manager must be an honest and objective individual and must impart this honesty and objectivity to others. He or she must be able to deal effectively with difficult people; honesty and leadership are the key attributes for this task.
9. An absolute dedication and discipline for developing, documenting, and adhering to a detailed project management and task-oriented scheduling system. It is also helpful to have prior knowledge and experience with the sequential constraints involved with scheduling skilled trades.
10. An ability to admit errors and move on to correct them. The project manager *should not* be a bullheaded individual who insists that his or her way is the only way. This is particularly important in layout work and materials handling systems design and implementation.

(Having listed the 10 major attributes of an ideal project management candidate, if any of you have *all* of them, please send the author a copy of your resume. You are one of an elite group.)

THE COMPUTER—AN INVALUABLE PROJECT MANAGEMENT TOOL

A computer-based project management system (CPMS) with resource leveling capabilities is a definite asset for major projects. Even a simple Gantt chart (see Figure 2-14) without resource leveling capabilities can be extremely helpful. Personal computer-based software systems are now used extensively.

CPMS software packages currently on the market include:

- Microsoft Project™, Microsoft Corporation
- Harvard Total Project Manager™, Software Publishing Corporation
- SuperProject™, Computer Associates
- Primavera Project Planner™, Primavera Systems
- MacProject™, Claris Corporation
- Fast Track Schedule™, AEC Software Incorporated.

Several of these programs have IBM PC and Macintosh versions.

In addition to project management, a PC or Mac also can be a great aid in helping to accumulate, record, analyze, and project basic facility planning data. This type of work is commonly referred to as *decision support* or, in total, as decision support systems (DSS).

DSS can be extremely helpful to the layout planner in the data gathering, initial analysis, project presentation, and project approval phases. These hardware/software-based routines are particularly useful in helping to project space needs and develop balanced operation process flows. Most of the computer routines used in this phase can be purchased in commercially-available software packages. They can be used for:

- Statistical tabulations and calculations;
- Curve fitting, extrapolation, and projection;
- Product quantity plotting/charting;
- Parts grouping and ranking;
- Process/materials flowcharting;
- Line balancing;
- Simulation;
- Linear programming;
- Business graphics;
- Financial modeling.

By far the most useful tool for DSS is a desktop PC running a standard spreadsheet program with good charting or plotting capability. A color printer or plotter is also useful. Computer-based layout programs are discussed in detail later in the book.

SETTING THE PATH TO SUCCESS — DEFINING THE ATTRIBUTES

Those who have had prior experience in developing facilities plans in general (and detail plant layouts in particular) know how harmful subjectivity can be in the final alternative evaluation phase. It is a humbling experience to have shotgun-type criticisms directed at plant layout alternatives at the end of what otherwise might have been an excellent analysis and approach. A weighted factor rating scheme should help prevent such problems.

Briefly, to make the best use of a weighted-factor rating system, it is recommended that most, if not all, of the scoring of success attributes be determined at the beginning of the analysis phase before any plant layouts have even been considered. It is also helpful to get upper management (those who will make the decision on the alternative to be selected) to help define and weigh the attributes at the beginning of the project. This is sometimes very difficult. However, if you are successful at the beginning, it will help make management feel they are part of the planning team. As long as there is frequent communication with management throughout the subsequent planning process, it will ease management's buy-in, preventing unwarranted shotgun criticisms at the end of the process.

Typical layout attributes that members of management like to compare between alternative layouts are:

- Flow-of-materials effectiveness,
- Flexibility for capacity expansion (or shrinkage),
- Cycle-time reduction,
- Overall reduction of WIP inventories,
- Materials handling cost savings and payback,
- Flexibility of physical change or rearrangement for new products/lines,
- Appearance or corporate image,
- State-of-the-art installation,
- Implementation costs.

This list is not all-inclusive. Your own situation may call for additional attributes. As an example, an entirely new site selection would involve a multitude of other attributes. Several other key evaluation factors are added to the list at the end of the process when we evaluate the various layout alternatives.

Generally, a weighting scale of 1 to 10 points is used. Each attribute is given a weight of 1 to 10 in importance by the management team. This is usually accomplished in a give-and-take meeting with management. Several attributes may have the same weight, but the meeting leader should strive to make sure there is some variety and spread in the weights. For instance, if a weight of 10 is assigned to all attributes, the exercise is meaningless.

You will usually find that management places a very strong emphasis on a few key items. Surprisingly, in many situations implementation cost *is not* the strongest attribute on the list.

The most important result of this exercise is to get the management team together and talking to one another about the project in the beginning phase. You need to pinpoint the main objectives (and subobjectives) that are driving the project. You also may be surprised by important hidden factors, subobjectives, and agendas that you might not have otherwise learned about until the end of the layout process. It is better by far to smoke out these items at the beginning than to face nasty surprises at the end.

A TYPICAL LAYOUT PLANNING APPROACH

Figure 3-1 is a graphical overview of the Manufacturing Plant Layout (MPL) process. You should review this step-by-step procedure outline and refer to it periodically. However, before we discuss the details of the procedure and the relationship logic required for the plant layout, we must address basic data needs and materials handling requirements. These are the first two task outlines in Figure 3-1, labeled "Data Collection" and "Material Flow Analysis." The Roman numeral numbering system is not meant to imply a required sequence (except that data collection starts the process). One can work procedures numbered I, II, III, VI, and VIII without ever working on the development of space needs in procedures IV, V, and VII.

All of the procedures shown in Figure 3-1 are discussed in the remaining chapters of the book.

From this point on, we assume that you have already selected the site (external boundaries), the building, or the specific area boundaries/limits of the overall area to be laid out.

A PRELUDE TO THE DATA GATHERING PROCESS

The first step in any layout process is compiling the necessary information for analysis. This step is labeled as "Data Collection" in the procedure outline.

Note that "Data Collection" appears to be the smallest block in Figure 3-1. It should be emphasized early on that the sizes of the blocks shown have absolutely no bearing whatsoever on the amount of work required. Data collec-

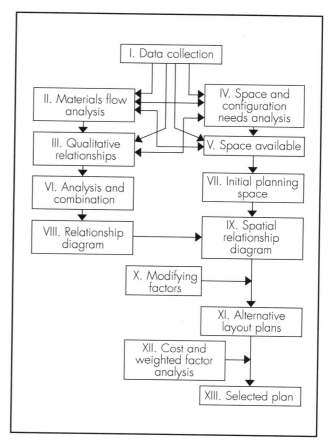

Figure 3-1. *This pattern of procedures incorporates all of the major steps in the journey to a sound macro plant layout plan.*

tion usually accounts for 50 to 60% of the work on a plant layout project. In fact, in my experience, the more time spent on data collection (and analysis) as a percentage of total effort, the better and more successful the plant layout alternatives are.

Obviously, there is a limit to the amount of time one can spend on data collection and analysis. However, if the approvers of the plant layout believe that shuffling machine and equipment templates around on a blank piece of paper defines the optimum layout process, they need to be educated. Likewise, if the approvers believe that the drafting of 20 plant layout alternatives (without an in-depth analysis) and several pot-shot critique sessions take most of the work effort, you and your company are in for a very rough experience.

It cannot be stressed enough: allowing a reasonable amount of time for data collection and analysis pays for itself hundreds of times over. Each layout project is different. For smaller uncomplicated projects of under, say, 20,000 to 40,000 ft² (1,850 to 3,700 m²) in area, 4 to 5 weeks for data collection (assuming no delays) should be adequate. For larger, relatively uncomplicated projects of, say, 375,000 to 650,000 ft² (35,000 to 60,000 m²), the planner should allow at least 10 to 14 weeks for data collection, assuming no delays. Obviously, if there are multiple business units housed under one roof, each with its own management group, manufacturing area, and receiving and shipping area, more time may be required.

PLANNING CONVENTIONS AND STANDARD SYMBOLS

Figure 3-2 summarizes the symbols and drawing conventions used for plant layout and facilities planning. The recommended colors are adapted from the International Materials Management Society's *Standard Color Codes for Use in Layout Planning and Materials Handling Analysis*. The process chart symbols are from an American Society of Mechanical Engineers (ASME) standard.

The activity area colors are particularly useful when attempting to consolidate similar functional activities within a plant, e.g., centralized support functions or centralized storage areas.

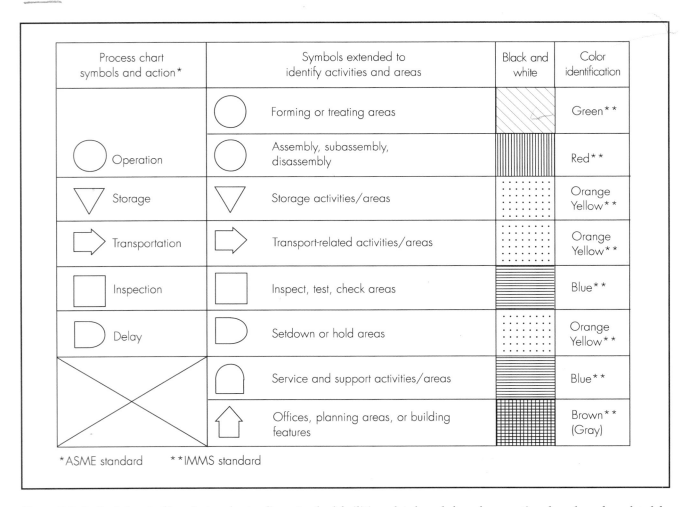

Process chart symbols and action*	Symbols extended to identify activities and areas	Black and white	Color identification
	Forming or treating areas		Green**
Operation	Assembly, subassembly, disassembly		Red**
Storage	Storage activities/areas		Orange Yellow**
Transportation	Transport-related activities/areas		Orange Yellow**
Inspection	Inspect, test, check areas		Blue**
Delay	Setdown or hold areas		Orange Yellow**
	Service and support activities/areas		Blue**
	Offices, planning areas, or building features		Brown** (Gray)

*ASME standard   **IMMS standard

Figure 3-2. *In the interest of broadest understanding, standard facilities-related symbols and conventions have been formulated for facilities/manufacturing engineers.*

4

BASIC DATA NEEDS

DATA REVIEW AND ANALYSES— FOUR MAJOR FACETS

When collecting data, the first thing we have to acknowledge is that most data have a "lifetime." That is, it normally has a future and a history as well. For most projected data, we should ask the following questions (where appropriate).

- Where are we coming from (the historical perspective)?
- Where are we now (today's conditions/parameters)?
- Where do we intend to go (future conditions/parameters)?
- And finally, we need to ask: How will we get there (physical implementation plans)?

For example, if we are to plan an expansion to produce, say, 200,000 units annually in 18 months, how many units were we producing 2 years ago? How many 5 years ago? How many units are we producing per year now? Is this projected increase attributable to a natural progression of increasing capacity brought by adding equipment?

Are we assuming a normal increase in productivity? If not, are we assuming a higher level of existing equipment utilization? If we are assuming a normal increase in productivity, how are we going to get there? What is physically going to change or improve to meet the target production numbers?

The same questions can be asked about projections for marketing plans, product configurations and models, inventory turnover, equipment utilization, etc. For example, if a company is moving toward cellular manufacturing, in many cases equipment utilization may be lower than before the change. The resulting increase in total equipment requirements has to be taken into account in the plant layouts.

Other similar important pieces of data are inventory turnover goals, particularly WIP inventories. Suppose the company is projecting an inventory turnover rate of 24 per year to coincide with the expansion completion 18 months from now. If inventory turns were 6 per year 5 years ago and 10 per year today, what magic potion is going to drive them to 24 in 18 months? What must change physically to meet these goals? Certainly cloning an existing layout will not work. Many of these questions will have a direct bearing on space requirements. All involved with the plan need to be made aware of the physical effects of changes in the way the company conducts its business.

CATEGORIZING DATA NEEDS

Know Your Ps and Qs to be SAFE

Minding one's Ps and Qs in merry old England meant looking after the number of pints and quarts of beer one was consuming. In our context, however, Ps and Qs take on a different meaning. Because specific data needs vary with difficult manufacturing processes and products, producing an efficient and effective layout is dependent on knowing the Ps and Qs of manufacturing plants.

Product: the desired end product of the manufacturing process, stated in terms of raw materials, piece parts, subassemblies, final assemblies, finished goods, etc.

Process: the activities directly required to produce the end product.

People: the people on the shop floor who must be involved in the planning and who ultimately make the processes and layouts work.

Quantity: the amount of each material needed or the number of parts or assemblies to be processed within a specific time frame, as well as the shift and seasonal variations in production requirements.

Quality: the tolerance standards and limits within each process, assembly, or product that are acceptable.

From this data, the designer can begin to determine the type of equipment needed and the level of sophistication required for each type of equipment.

Be SAFE

In addition to the basic *Ps* and *Qs*, other fundamental data must be added to help develop a layout plan.

Space: the size and configuration of the area available or projected for use.

Activities: the total activities necessary to produce the process, including all support activities such as battery charging, maintenance, control functions, packaging, etc.

Flow: the process order or routing in which these activities must occur to provide an efficient production process.

Equipment: the equipment needs calculated by analyzing the *Ps* and *Qs* above.

The O, P, Q, R, S, T Approach

When all data needs and activity relationships are considered in the selection of an optimum plant design, the layout stands as an efficient and effective tool for the company, as well as a representation of a safe environment in which the human assets of the company can produce. Similar to the above *Ps* and *Qs*, in the 1960s and 1970s, pioneering industrial engineers such as Muther* attempted to summarize all of the data

needs into easily-remembered mnemonics such as O, P, Q, R, S, and T ("O" has been added by me). Actually, data requirements for developing today's plant layouts are not quite this simplistic, but Muther's original work is still valid as a training tool. We can broadly define the O, P, Q, R, S, and T data needs as:

Orders: the characteristics, sizes, and frequency of customer orders, scheduled due dates, the variance of order sizes, the physical cube and lot sizes (of both internal and external receipts and shipments), etc.

Products: the various classifications by physical characteristics and use, model number, type of ordering, customer service requirements, product mix, special order or control, etc.

Quantities: the production quantities (per consistent unit of time) of the various products or parts produced and the rationale used to establish lot sizes or run lengths.

Routings: the shop routings (typically of a selected number of important products or parts) and the material flowpaths. Data is needed on the configuration of the handling containers used to transport parts and materials and on each piece of handling equipment (fork truck, roller conveyor, etc.) used in the movement process.

Support: tooling, maintenance, inspection, contract services, etc.

Services: break/lunch areas, shop offices, restroom requirements, etc. Even though no flow of materials may exist between these functions and the production shops, we need to account for the required relationships and the required area set-asides.

Timing: number of shifts per day and per year on which production levels are based, seasonality effects, timing of inventory buildups and drawdowns, lead times of subcontracted parts and supplies, scheduled plant shutdowns, maintenance downtime, etc.

These mnemonics should help you remember the basic data needs. The next section expands upon these basics.

---

*Much of today's recognized work in plant layout and facilities planning can be traced to Richard Muther, a pioneer in systematic layout planning.

FUNDAMENTAL DATA

What types of data are key to properly planning a plant layout? The following short list includes the most common data needs.

- Product/part configurations.
- Short- and long-term production quantity/volume projections for each product/part configuration (or a representative sample of production parts).
- Production activities listing with existing and/or proposed cell/departmental definitions and current and proposed space assignments. The activities list also should include general or common activities such as lunchrooms, mechanical rooms, etc., that will occupy a portion of the area to be laid out.
- Operations process charts showing labor and equipment cycle times and sequential routings.
- Information on current materials handling (MH) techniques, along with quantities and configurations of current handling containers and MH equipment.
- Timing requirements (scheduling, number of shifts, inventory buildup and drawdown periods, and seasonal and cyclical variations).
- Supporting service requirements: those support activities and space needs not covered in the production activities listing.
- Inventory turns: history and plan (including raw materials, WIP materials, finished goods, packaging materials, etc.).
- Equipment utilization: historical and plan.
- Operating personnel interviews.
- Environmental regulations affecting operations.
- Existing plant layout to scale (if available) showing all load-bearing walls and building column locations.
- Total space requirements or constraints (if known).
- Production equipment listing and scaled drawings of each piece.
- Production labor standards (historical, engineered, or measured).
- Information on order characteristics (lot sizes, physical cube sizes, delivery timing, expected variances, and changes).
- Information on expected materials receipt schedules and storage philosophy (JIT, daily, weekly, monthly, *kanban*; point-of-use storage or centralized storage; etc.).
- Existing manpower, direct and indirect, by shift and classification, along with associated labor and fringe costs.
- Listing and configuration of "monuments."
- Management philosophy, vision, success attributes, etc.

Monuments

Although most of these key data items are somewhat self-explanatory, "monuments" need further explanation. A monument is a piece of equipment, a particular function or installation, or perhaps even a whole department that upper management has deemed *will stay in its current place forever*. No matter how good a layout could be, if the monument was moved, the cost of moving it would be too high in upper management's view.

Keep in mind that the term *cost* may include considerations other than pure monetary cost. I worked on one project where the client company's founder's original laboratory (garage) was located in a section of the site. The sentimental value of the garage prevented its being moved, severely hampering an optimum growth and expansion pattern. A similar situation occurred with a company-sponsored apple orchard that prevented an optimum facility growth pattern. That plot had been started by the company's current chairman and the entire management team could not bring themselves to recommend its removal; it had literally become sacred in their eyes.

Other monument costs might involve reports or negotiations with governmental regulators if land with an unknown history is disturbed (e.g., exposing underground hazardous materials). Managers of aging chemical plants who do not have clear history of their site's usage are sometimes reluctant to do *any* kind of digging on their plant sites.

Reverse monuments

Similarly there have been cases of what I term *reverse monuments*. Usually, but not always, *reverse monuments* are surprises and are not known by anyone in advance. It brings to mind one rather humorous case involving a very large injection molding plant.

The plant had been operating for years on a rural site (they had their machines and processes in place before stringent environmental regulations were in effect). The company drew their mold cooling water from an underground aquifer. The relatively clean water was treated both before and after the cooling process, and it was allowed to drain through a large pipe into a swale after use and treatment. The swale formed a small stream that exited the property. Since the water leaving the plant was purer than when it was pumped from the company's well, there did not appear to be any problem. However, the small town closest to the plant was growing in the direction of the plant. The company was getting mild community pressure and local farmer pressure that they really should be re-using their water through a closed-loop water system and not "draining" the aquifer. As part of the plant expansion, and in the spirit of being neighborly, the company designed and installed a closed-loop water system.

Unknown to anyone at the time, the water that had previously left the property made up more than 60% of the small stream's volume. That stream in turn fed a small swampy area some distance away which contained an endangered species of tadpoles and frogs. The company had literally stumbled upon a *reverse monument*. In attempting to be a good neighbor and do the right thing, they had done precisely the wrong thing in the eyes of the environmentalists. I do not know what the cost was in time and money for legal fees or if the situation was ever resolved to the satisfaction of the environmentalists. The company's closed-loop water system was installed. Presumably, the taxpayers paid to relocate the frogs and tadpoles.

ISSUES AND PROBLEMS IN DATA COLLECTION

The data needed for analysis is not necessarily easy to acquire. In some cases, only fragments of the information are available and the planner needs to use his experienced judgment to fill in the blanks.

Among the issues that must be resolved during data collection, analysis, and evaluation are:

- The degree of accuracy and completeness of current and historic data.
- The probability of the accuracy of projected production quantities and other items that need to be forecast. The farther into the future the projections go, the more likely they will be inaccurate.
- The degree of flexibility needed to accommodate change.
- In the early stages of the planning process, the degree of specificity required for production efficiency.
- The optimal relationships between the activities/functions.
- The weights assigned to distribution, production, operating cost, and initial cost factors, and to conflicting company objectives.
- The degree of mechanization/automation that can be justified.

Resolving Accuracy Problems

The accuracy check should include comparisons of data for identical and similar time frames gathered from different internal or external sources. Data from several time frames should be compared to assure that predictable peaks and slumps in production demand are identified and that the degree of uncontrollable or unpredictable fluctuation in production is estimated as accurately as possible.

When holes exist, some data can be estimated from known relationships among available data. Other questions about accuracy can and should be resolved by the person in charge of the area from which the data is gathered.

All data must be converted to compatible units and time frames, of course. Missing figures that cannot be calculated using available data must be estimated from experience.

Accuracy of Projected Volume and Needs

Designing a facility or plant layout requires information about expected production volumes, trends, and predictability of future demands for products. The less specific the product, process, and schedule data, the more general-purpose the facility plan must be. This is typical for high-tech industries.

The more specific the input data about product, process, and schedule designs, the greater the likelihood of designing the optimum layout or facility to meet *short-term* company objectives, not necessarily for long-term results. Asking for the marketing department's help in explaining the reasons for predicted trends as well as for volume/quantity estimates can prove helpful in estimating the accuracy of projected figures. However, the planner needs to be wary of the eternal optimism usually expressed by marketing departments. It is recommended that three different scenarios be developed when planning a new plant or major rearrangement—*optimistic, pessimistic,* and *most likely.* We discuss each in a later chapter.

Estimating the Flexibility Required

During early design stages, general data suffices, and general data—hence flexible arrangements—provide adequate information for early evaluation. The degree of flexibility incorporated into the final layout depends on the reliability of projections, the stability or life cycle of the product design, and the value management places on the ability to move quickly into new or redesigned products. This value should be stated in the company's list of objectives.

If increased flexibility can be attained without sacrificing production efficiency, flexibility ought to be incorporated into the layout design, whether or not objectives directly state the need for flexibility. Clearly, those operations that are most likely to expand in the future should not be boxed into a corner or against a wall that cannot be altered or moved at a later time.

The Degree of Specificity

The degree of specificity often increases with the degree of mechanization or automation incorporated into the production process. Basing the handling system on standard handling units such as pallets of a specific size, uniformly-sized

containers, or totes often helps retain flexibility without compromising mechanization or automation possibilities. The planner should delay detailing until the final stages of the layout process unless there is potential for major changes in process or handling equipment.

Establishing Optimal Relationships

Methods for establishing optimal relationships are built into the manufacturing plant layout (MPL) process. These are discussed in the following pages. Here, it is sufficient to say that conflicts among relationship priorities often can be solved by providing alternatives to direct proximity. These can include alternate methods of communication, such as computer, phone, or radio; alternate process sequences; and alternate storage, distribution, or location arrangements. None of these alternatives should be discarded until they are evaluated in relation to the proximity needs of all functions. Only by remaining open to all options can the designer choose the best alternatives to satisfy all functional needs efficiently.

Weighting Factors for Use in Evaluation

While the designer must evaluate the alternative layouts, evaluation criteria should obviously be checked with management. Selling management on the recommended layout often depends on understanding the real value management gives to the ability of the layout to meet management objectives. In a highly competitive market, speed and production costs often are primary factors. If rapid model changes are needed to keep pace or capture a greater percentage of that highly competitive market, flexibility and speed may become more important than production costs unless costs are radically higher. Therefore, reduced cycle times for new product introductions may carry a heavier weight than pure production costs.

It often proves more expeditious to put these evaluation criteria or success attributes into the order you feel reflects management goals before presenting them to management for approval. Weighted factor evaluations are discussed in Chapter 13.

The Degree of Mechanization/Automation

Mechanization and automation considerations are an integral part of industrial upgrading for efficiency in America and in Japan. Much of Europe's industry is also highly mechanized and automated. However, automation remains extremely expensive and difficult to justify for many small and medium-sized manufacturing concerns. The MPL weighted factor evaluation method often provides a way of helping to justify automation and mechanization on factors other than pure costs (e.g., strategic goals, etc.).

An important factor in providing the best possible design remains keeping an open mind. In this process, the designer must strike a balance between the input from quantitative analysis of data, the physical requirements of equipment and handling, and the more subtle factors of management policy, marketing goals, and finan-

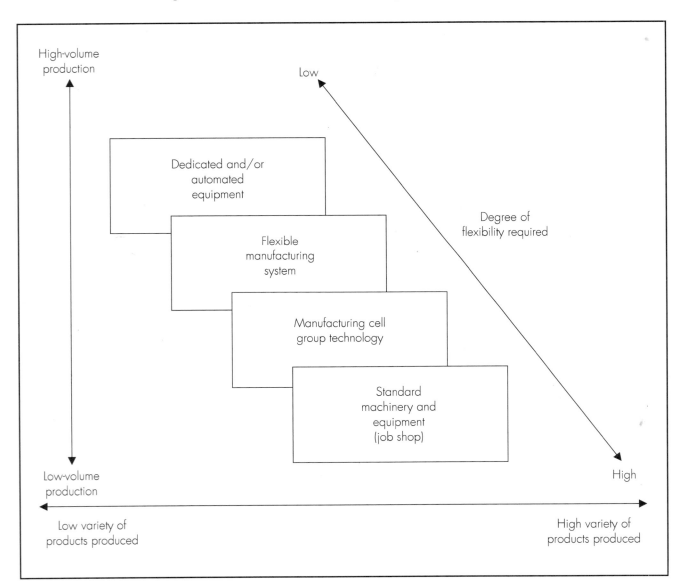

Figure 4-1. *Product focus is the key determinant of production system type. High-quantity, low-variety products call for dedicated equipment in a product-focused layout; low-quantity, high-variety product offerings require the flexibility of universal types of equipment laid out in a traditional process-focused layout.*

cial requirements. Closing any option too early is a mistake.

The Manufacturing Spectrum

An overall global view of what I call "the manufacturing spectrum" is shown in Figure 4-1. The figure depicts the overall spectrum of manufacturing activities within the manufacturing "universe." On the vertical scale are the production quantity levels, varying from very low to very high. The horizontal scale axis depicts the potential wide variety of products produced. As the graphic shows, the variety of products produced also may vary from very low to very high.

High-volume, low-variety production

As we move along this continuous spectrum, the degree of flexibility required for our production system also changes. As an example, let's assume we are manufacturing razor blades or cigarettes. The annual quantities for these types of products would be very high while the variety produced would be relatively low. This type of high-volume, low-variety production would probably fall at the upper left corner of our spectrum—dedicated, automated, custom-built machinery would be used for production. The processes, equipment, and production routing would be fixed and relatively inflexible for changes. An in-line, product-oriented production line type of layout would probably be used with little, if any, human handling of in-process materials (other than supplying raw materials and packaging materials to the line).

Low-volume, high-variety production

Let's now look at the other end of the spectrum. In a pure job-shop, custom-build type of production operation, one never knows the details of exactly what needs to be manufactured far in advance. Typically, this type of operation may build one-of-a-kind products or serve a very large customer base by manufacturing small quantities of their "odd-ball" products. A typical "pure" job shop produces a great variety of parts or products with relatively small quantities of each. Due to the high variety of customer needs, the job shop must maintain a very high degree of flexibility in processes, equipment, and production routings. Typically, there is also a high degree of human intervention with in-process materials, usually in the physical movement of materials and parts from one operation to another. This type of operation would be shown somewhere near the lower right-hand corner of the spectrum shown in Figure 4-1. The job shop tends to have universal types of equipment laid out in a traditional process-oriented layout. In a job-shop type of machine shop, for example, there would normally be a dedicated area for lathes, another separate area for milling machines, another area for grinding machines, and so on.

These two ends of the spectrum offer the most efficient operations for the types of production described. However, the vast majority of industrial operations usually have a mix of production needs. Some of the production requirements generally fall between these two extremes. In the mid-range areas, product-focused manufacturing cells or computer-controlled flexible manufacturing systems (FMS) should be considered.

Flexible manufacturing systems

Computer-controlled FMS have been used since the 1960s by very large manufacturers—typically automotive and defense equipment manufacturers—to replace transfer (machining) lines. Although pioneered in the U.K. and U.S. by companies such as Molins and John Deere, respectively, they have been much more popular in Europe and Japan. FMS are very expensive strategic investments (I have never seen an FMS justified on pure labor savings alone). They are usually suited for only very large companies which have the technological base in place to support them. Also, there has been a phenomenal increase in computer capabilities over the last two decades. Coupling increases in computer hardware and software technology with cost reductions and increases in productivity and accuracies of multipurpose specialty machines has led to less expensive investment alternatives to FMS. As a result, many older FMS are now looked upon as being somewhat less "flexible" than newer lower-cost alternatives. Many of the "strategic" FMS versions that are more than 5 years old are now operating with obsolete general-purpose machines running on aging minicomputer-based software/hardware systems. Strategic FMS investments have not always kept pace with advances in technology. In many situations, FMS have become fixed-in-place monuments. Although still highly productive, many older systems are now too costly to update, leaving their owners somewhat behind the technological curve, thus hampering potential improvements in plant layout. Keep in mind, however, that FMS did pioneer a single-piece flow concept and continues to reduce the economic batch size to one piece. However, for most companies of a lesser size than the automotive and large defense contractors, manufacturing cells or flexible manufacturing cells (FMC as they are frequently referred to) are a better choice.

Between FMS and the job-shop portions of the manufacturing spectrum of Figure 4-1 is an area suitable for manufacturing cells. Cells and pull-versus-push systems in general are extremely important facets of plant layout.

The manufacturing spectrum spawns an important *variety-versus-volume* concept that helps us determine the most efficient production methods and plant layout(s) for our operations. Usually based on a smaller scale but covering similar concepts is the Product/Quantity Chart. That, too, can help us decide on the most efficient types of production methods to use.

PRODUCT VERSUS QUANTITY

At this point in our procedure, we have collected a volume of historical, current, and projected production information. We have also briefly discussed the wide spectrum of potential manufacturing activities. We now need to review the production quantity requirements and how those requirements may affect our manufacturing philosophy, production methods, and plant layout plans.

P/Q Analysis

One of the tools in the planner's toolbox should be the product/quantity (P/Q) analysis. Although the analysis does not fit nor will it benefit all companies, it is worthwhile to know about. This data — products and quantities to be produced in a given period of time — are often

provided by the marketing department, but should be double-checked with engineering and operating personnel. It is important to recognize that time is a major consideration of the P/Q chart. When developing our charts, we always want to be working with a consistent time period; usually, but not always, *annual* quantity requirements are used. Some companies with highly seasonal production will need to use a shorter time period. Obviously, projections of future needs also must be evaluated along with current and historical data.

P/Q Charts

Displaying the annual quantity required of each product in a chart shows those products that are fast movers and those that are slow movers. An example P/Q chart is shown in Figure 4-2. The curve produced in such charts often reflects "the 80/20 rule" — where 80% of the annual quantities of goods manufactured involves only 20% of the number of products produced.

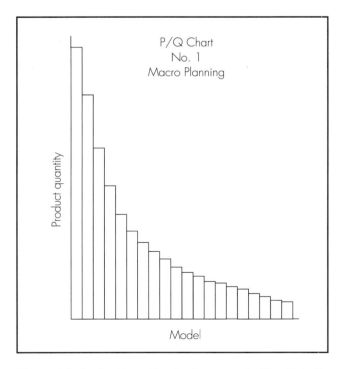

Figure 4-2. Product/quantity charts are a valuable aid to the layout planner in determining the type of production and layout arrangement.

But all mixes are possible. For many plants, one end of the curve usually shows large quantities of relatively few products or varieties, while the other shows many different products, each requiring small quantities. The chart axes are not unlike the axes on the manufacturing spectrum graphic discussed previously.

For the layout planner, the product/quantity curve has special significance. It forms a part of the basis for determining the fundamental type of production and layout arrangement; e.g., mechanized/automated operation, assembly-line production, manufacturing cell, job-shop, combination, or split arrangements of two or more schemes, etc.

Keep in mind that there may be an entire series of P/Q charts for a particular plant. Normally there is at least one overall chart for the top-level product models or assemblies being produced. However, in a pure job shop, there frequently are no pure "products" being produced. The job shop may only produce parts or components of someone else's product. However, most companies, including job shops, will still have a P/Q chart; one which shows all of the *individual piece parts* being produced (instead of or in addition to product models). If there are major product subdivisions or business units that will remain as separate entities within the plant, they will have their own P/Q chart or several charts.

Why should the planner be interested in the P/Q chart? Suppose we are producing 500,000 of a particular model annually (shown at one end of the curve on the left-hand side of Figure 4-2) and only 5,000 annually of another model (shown at the other end of the curve on the right-hand side). Would you plan to use the same kind of equipment and layout to produce these two different models? Would you use the same production or assembly line? Would you use the same shop supervisor and the same work crew for both models? Obviously the answer depends on your own particular situation, but in most cases the answers to these questions would be no, as we see in the next section. Similar to our discussion of the manufacturing spectrum, different variety/volume (P/Q) combinations can be more efficiently produced with different production methods and different plant layout schemes.

The planner also should be aware of some of the major pitfalls that can occur when blindly comparing part or product quantities. Quantity alone does not always give a true picture of importance or variety. For example, a manufacturer may produce 50,000 small washers a month of the same size and 500 precision (and different) herringbone gears a month. Concentrating on quantity alone may be a misapplication of both importance and resources. Frequently, companies also chart *gross profit contribution* and *direct labor* per part or product in addition to sales revenues and total production quantities. From a business perspective, these are frequently much better indicators of importance. They also help highlight where a company's manufacturing and industrial engineering resources should be focused.

Potential Splits and Combines

Splitting or combining operations can offer tremendous paybacks for many companies.

Deep P/Q curves, like the one in Figure 4-2, indicate that dividing product and production areas into different layout and handling systems may be the best and most efficient approach. Production of 15 or 25% of the total products (represented at the top left of the curve), that may account for most of the plant's production (remember the 80/20 rule), could be automated or mechanized if the quantity requirements so justify. Of course, as highlighted in the last section, when we are discussing the few products on the left of the chart, we mean *important* products, not hardware items such as rivets and screws (unless the manufacture of rivets and screws *is* our major business, which in that case would make them extremely important). When we look at deep curves, we should consider splitting the operations into at least two categories to achieve manufacturing efficiencies. Let's discuss an example.

I led an in-depth operations study of a manufacturer of electric and gas-fired water heaters. The products were produced principally for the consumer household market. The heaters were typically of the 40- to 80-gallon types that are commonly found in homes and apartments. The company also made some very small water heaters for recreational vehicles and pleasure boats.

The company was experiencing some minor product costing and pricing problems in relation to some of its competitors and its margins were being slightly eroded. The main problem, however, was that the company was having trouble meeting total customer demand. Everyone realized that they were producing mainly for the replacement market. If their company's product was not available, a consumer without hot water could not wait for a back-order and would instead go to the nearest competitor or plumber who did have the correct model in stock. Management believed everyone in the factory was working hard but the company had to improve productivity and lower its costs, particularly on its household product models.

In addition to a steel-tank and sheet-metal fabrication plant, the company had one major assembly line running on two shifts and a partial third shift. Even if the company could find willing workers (which they could not), the demand problem could not justify the addition of a full third shift. The company was forced to work overtime on the weekends to meet the demand requirements.

The assembly line was achieving only approximately 70% of its potential nonstop capacity. During the course of the project (which included a delay study to determine the significant reasons for lost production capacity) it was discovered that one of the major causes of lost capacity was the lost time encountered during model changeovers and parts "starvation." These model changeovers occurred quite frequently. At times the line also would be "starved" for parts and would need to shut down until parts could be expedited to the right places on the line. That is to say, the proper parts and subassemblies were not always located on the line at the right times.

Numerous "fixes" were required to optimize line productivity and keep up with demand. The one of most interest here was the addition of a new "short-run" line to handle the heaters for the boat and recreational vehicle markets. Splitting that high-variety/low-volume production

off the high-volume line allowed the main line's production to increase more than 26%. Better yet, this was accomplished with only a very small (less than 5%) increase in the number of employees! The main line was rebalanced to free up operators for the smaller, custom, low-volume line. High-cost overtime was no longer required to meet demand. Management had never realized the disproportionate amount of overhead and burden the main line was absorbing (and hiding) when producing the numerous models of small heaters. Neither had they known about the materials handling logistics and parts starvation problems caused by running too many low-quantity models on one line mixed with high-quantity models. In essence, the company was trying to produce all of the models shown on their P/Q chart on one assembly line.

Again, there is a lesson to be learned here. If we have a deep curve on our P/Q chart, we should consider splitting our operations into at least two categories to achieve manufacturing efficiencies. Similarly, suppose we are manufacturing and assembling, say, 3,500 color printers per day of one model, and 100, 75, and 50 of three other models. Should they all be assembled on the same line? Usually not. The inefficiencies of producing low-volume items on the high-volume line imposes unnecessarily high costs on the company's main line of products. This also tends to hide the true costs of producing marginal products.

By analyzing the P/Q charts of several products, we can sometimes see the reverse situation. For example, if we have several relatively shallow P/Q charts and we combine them (notwithstanding the politics associated with combining several different business units), significant benefits may accrue. We can sometimes make a case for combining operations to allow the implementation of a high-volume, high-productivity line that is common to several other product lines (see Figure 4-3).

Frequently, the mid-range area of a P/Q chart offers opportunities for implementing focused manufacturing plants or work cells. Typically, the required production quantities are too low to justify full automation or mechanization, and the product variety is also low enough to make

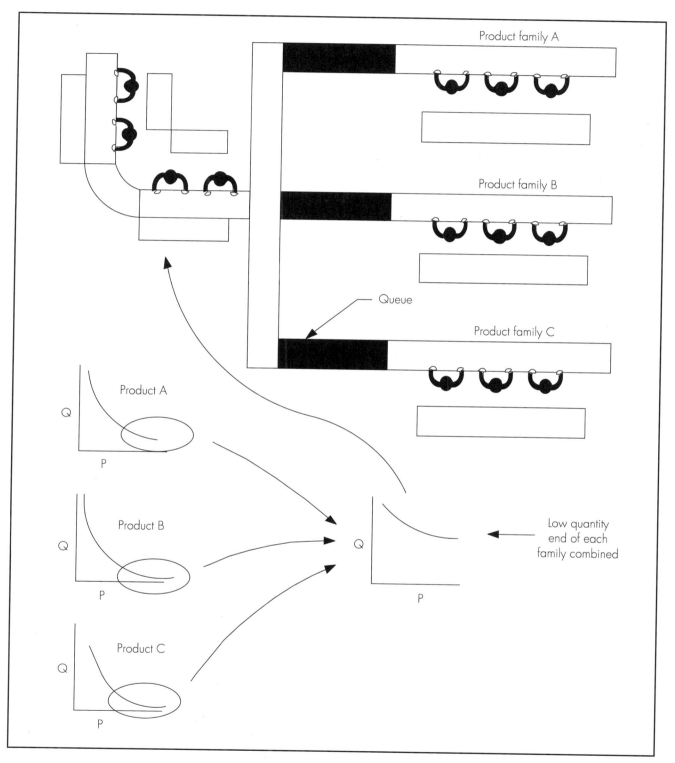

Figure 4-3. *In some cases, combining several low-volume operations in a common area will produce efficiencies of scale by feeding separate high-volume, high-productivity lines products requiring similar production operations.*

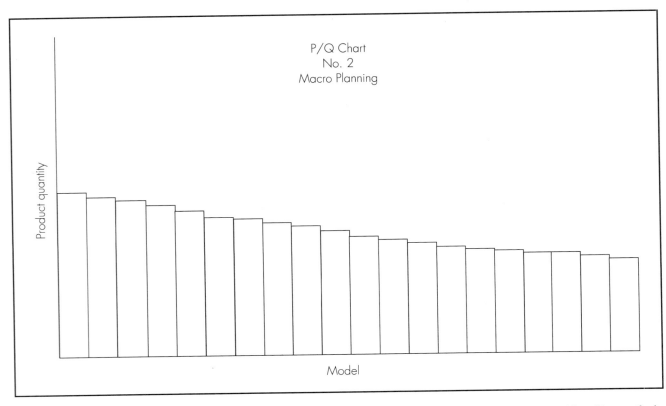

Figure 4-4. Shallow P/Q curves indicate typical job-shop operations, with traditional, process-based layout and handling methods.

a traditional, batch-oriented process inefficient. (We discuss manufacturing cells and "pull" manufacturing systems in general in the next section.)

Products with characteristics that fall within the far right range or shallow curve range of the P/Q chart are typically best suited for handling by universal equipment using traditional, fixed-location processing areas. The shallow curve shown in Figure 4-4 could be a total company P/Q curve; this would be the case with a typical job-shop or custom-build operation. It also could be viewed as the right portion or subset of a larger curve. This type of low-quantity, high-variety production normally calls for a traditional, function, or process-based layout and universal handling methods. The layout plan should center the greatest efficiency on those items in the center of the curve, rather than the few items on either end. Of course there are exceptions to every rule. Notwithstanding the water heater example, many low-volume items with similar physical characteristics are fabricated or assembled in manufacturing cells. Usually these manufacturing cells are specifically designed to handle "families" of similar products or parts.

A typical batch-oriented layout is shown in Figure 4-5. While in Figure 4-6 a typical layout with a combination of slow movers and high movers is shown. Manufacturing cell layouts are covered later in the book.

PROCESS VERSUS PRODUCT

Figure 4-7(a) shows a simplified schematic of a product-oriented layout. This is usually how a small company or product group within a larger company might start its manufacturing operations. Although it shows a machining operation, the concept would be roughly the same for almost any type of plant; there is nothing particularly unusual about the layout. Let's assume that the operation runs as shown for a period of time and eventually reaches its maximum capacity. We will also assume that the

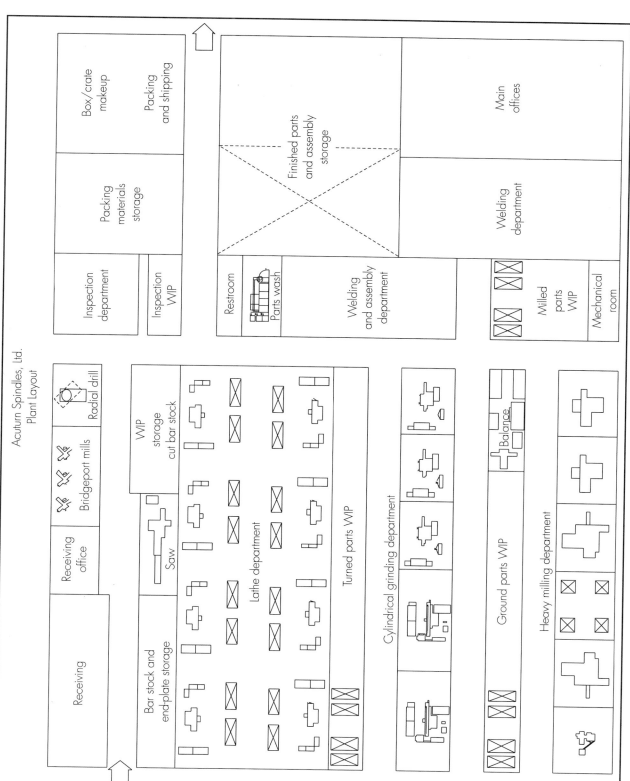

Figure 4-5. *A batch-oriented layout features methods, equipment, and flow to accommodate a process-focused philosophy.*

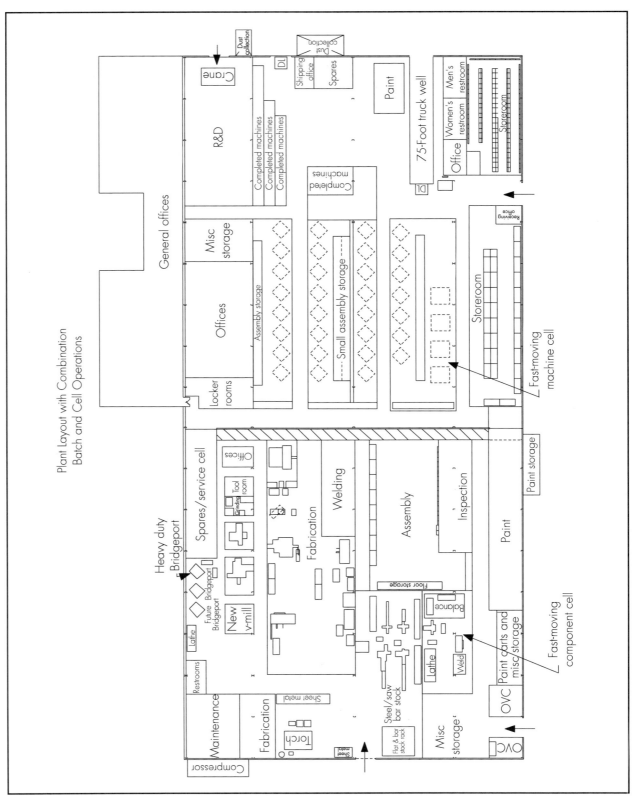

Figure 4-6. *To accommodate product lines of both low and high volumes, combination batch and cell layouts provide a viable solution.*

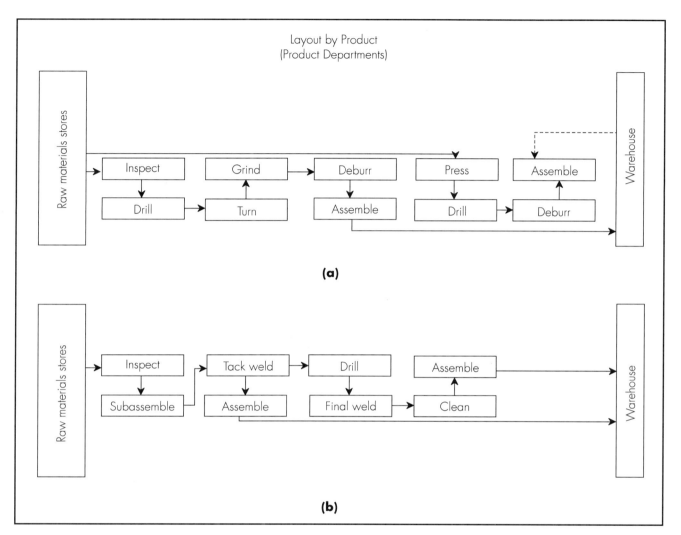

Figure 4-7. A typical product-focused layout.

company has developed another product with some similar operations and that it plans to install a new line to handle it, as shown in Figure 4-7(b). When we look at the two schematics, we can see that there are similar operations within each product-oriented layout.

If we foresee several more such products being developed, we might ask if we can develop manufacturing economies of scale by combining like operations. This is the typical progression for many growing operations. If we do in fact combine operations, we will likely get a schematic layout somewhat similar to the one shown in Figure 4-8.

Although the combined layout looks efficient by virtue of allowing specialization of labor and economies of scale, it generates some difficulties or tradeoffs as well. The principal problem is that the combined layout tends to "breed" inventory between every departmental function. If every operation were perfectly balanced and could complete the same size batch in the same period of time, this inventory buildup would not occur. However, in my experience, that kind of balance is almost impossible to achieve (see Figure 4-9).

On the plus side, line shutdowns may decrease with the combined layout since dependency is

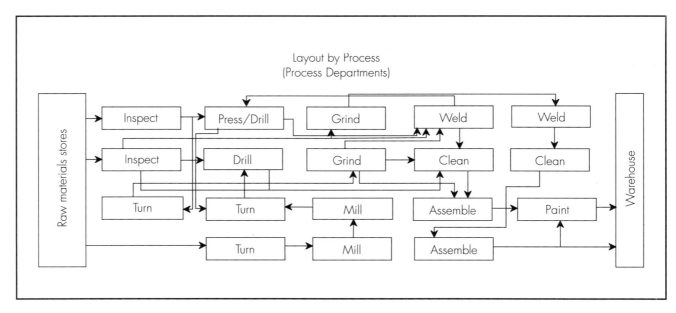

Figure 4-8. *Economies of scale can be achieved through a layout based on process, but certain tradeoffs become inevitable.*

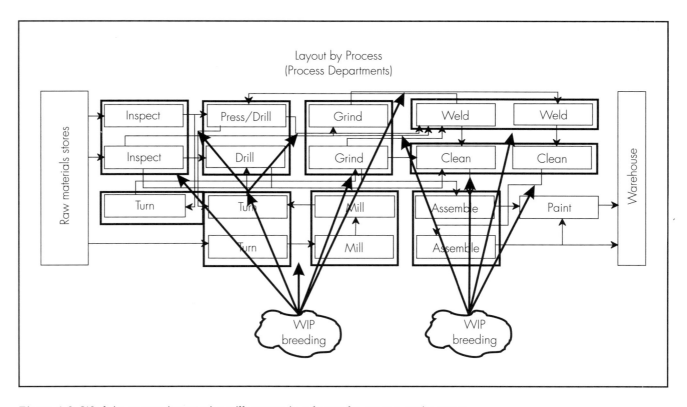

Figure 4-9. *Work-in-process inventories will emerge in a layout-by-process environment.*

reduced and covered (hidden?) by inventory buildups. However, on the negative side, cycle times through the shop may increase and quality levels may decrease with the combined operation. Instead of having two supervisors or lead persons with the product-focused layout, we may now need more lead persons with the process-focused layout. There also will be a very significant change in the way the shop is scheduled.

This discussion prompts some basic questions on manufacturing philosophy. Just what are we trying to optimize? Are manufacturing cells better than traditional centralized process types of arrangements, or vice versa? Before we can intelligently discuss the pros and cons of each, we need to discuss *push* versus *pull* manufacturing philosophies.

SCHEDULED PUSH VERSUS DEMAND PULL

In the previous sections, we discussed briefly several different modes of manufacturing and layout. In our P/Q discussions, these different operational modes were based principally on production quantity requirements. Many more factors other than quantity ultimately contribute to the determination of the final plant layout. In the last section, we discussed several of those factors—specifically, product- versus process-oriented layouts. One of the major determining factors for our plant layout is based on whether or not the company uses a scheduled-push or demand-pull manufacturing philosophy.

Scheduled-push Systems

Scheduled-push (SP) systems are typically driven from a computer-based production scheduling system. Such software/hardware scheduling tools are generically referred to as MRP (manufacturing requirements/resource planning) systems. They typically forecast upcoming production requirements using "smoothed" data or rolling averages of historical production and shipping information adjusted for any known exception changes. Normally, MRP-based systems have a tendency to build product to an inventory stocking plan. Fluctuations in demand are normally handled by buildups and drawdowns of inventory throughout the enterprise. In addition to working from historical data, these types of SP or push scheduling systems tend to consolidate and batch new incoming orders to obtain the maximum build quantities of a particular part within a given time frame (or *bucket* as the time frames are normally referred to). A week's bucket of time is a typical shop order release cycle time for an SP system. That is to say, a shop order for a quantity of any particular part is normally issued only once a week. Since SP systems build to an inventory stocking plan, the weekly quantity ordered of any particular part may have no relationship whatsoever to the actual instantaneous quantity demand for the part. Theoretically, SP attempts to level-load factory labor by letting inventory levels fluctuate up and down against a predetermined target. SP systems favor traditional, batch-oriented, process types of layouts where inventories are a key component used to help level-load the factory. In essence, an unmodified SP approach favors different departments performing different operations on batches of WIP inventories.

Demand-pull Systems

The demand-pull or DP system is geared to scheduling and producing only that amount of product that has actually been ordered (or demanded) by customers. Theoretically, the DP system strives to bring inventories to zero or at least down to an absolute minimum. It favors a just-in-time approach of materials receipts from both internal and external suppliers. To accomplish this, a "straight-through," demand-pull type of materials handling, production, and layout scheme is required. DP favors a single piecepart flow in a product manufacturing cell as opposed to the batch-quantity, departmental approach favored in the SP system.

To summarize the effects of these two different approaches on plant layouts, we can list the pros and cons of a pull-type approach using a manufacturing cell arrangement versus a traditional departmentalized, centralized process layout. Except where noted, the pros or positives of a manufacturing cell are the cons or negatives associated with the traditional layout approach for a production area.

The Pros of a Pull System

In the context of a focused, product-oriented, manufacturing cell, the pull system boasts these positives.

- Very low WIP inventories.
- Low shop cycle time—from receipt of order to first shipment.
- Requires less space because of reduced inventory storage.
- Fosters a spirit of employee ownership.
- Fosters a focused team spirit and promotes in-cell communications.
- Has the tendency to produce a higher quality output.
- Simplifies scheduling of several product cells when compared to scheduling a multitude of individual process departments.
- Records less stock loss (less mislocated stock) when compared to controlling WIP inventories with push systems.
- Records less stock loss or damage from excessive materials handling.
- Generally results in leaner management bureaucracies as a result of product-focused cells.
- Provides improved customer relations and shortened and improved lines of communication and feedback. This usually results from customer requests to have their suppliers set up a focused cell arrangement solely dedicated to their particular products.

The Cons of a Pull System

In the context of a focused, product-oriented manufacturing cell, the pull system has its drawbacks as well.

- Unless a very good balance is achieved, multifunctional cells generally have lower equipment utilization (except for the bottleneck piece of equipment which may be running at capacity).
- Unless there is built-in redundancy, when one piece of equipment breaks down, the entire cell may be shut down.
- There is an inherent loss of centralized technical expertise.
- Cells generally cost more to implement than a traditional layout.
- Cells may be somewhat inflexible with regard to change. With product-focused cells, it is frequently more difficult and costly to prototype and introduce new products which are not a "fit" with an existing cell. The introduction normally requires a whole new cell.
- Products generally have higher labor content if frequent setup changes are required (setup times/costs are amortized over a smaller batch size than in process-focused departments).
- Cells with high technical or skills content usually require a more flexible work force, additional training, etc. Instead of being masters at one trade or operation, workers must become cross-trained in several operations (this also may be considered a positive for many companies). Highly skilled people are frequently not comfortable doing multiple jobs.
- Materials are often received on a JIT basis. This frequently results in a need for more space in transactions-related areas such as receiving docks (e.g., many more deliveries, etc.).

Which Is Best?

Though most of these pros and cons are self-explanatory, some items need a bit more clarification. Cells tend to produce a spirit of employee ownership because the cell "team" can usually *see and touch the total product* being finished within the cell—they can immediately see the results of their efforts. Each worker within the cell can communicate problems very quickly and resolve them within the cell very quickly. In traditional, departmentalized arrangements, employees often see only their small piece of the product. They usually do not have a good feel for how their small operation affects the big picture; similarly, defects are discovered much more quickly when a cell is working with single-piece flow or very small batch sizes. Large, centralized or departmentalized operations working with larger lot sizes can fill an entire pipeline of part inventories before a defective part is found. This can result in much higher scrap levels and missed shipments. Thus cells tend to have higher-quality output.

Generally speaking, the pros for cells far outweigh the cons. True, there are a lot of negatives associated with manufacturing cells, but most of the negative impacts can be reduced or eliminated. It takes some innovative industrial and manufacturing engineering work, but it is well worth the effort.

It is also true, however, that in plants with highly skilled workers such as in ultra-precision component machining, products produced in *well-run* departmentalized shops tend to have lower labor content and higher productivity than cell arrangements. Typically this is true because the skill levels are built up empirically from intense experience in operating a particular type of machine. For instance a machinist who has operated a group of precision cylindrical grinding machines for 10 years may never be, nor want to be, proficient at running milling machines on a very part-time basis.

Also, we should try not to lose sight of some of the benefits of departmentalized operations. Centralization of technical skills can be extremely important in process and capital-intensive operations. Capital-intensive operations such as die casting, multistep cleaning, top-quality painting, heat treating, etc., are usually much more efficient when they are centralized and not dispersed throughout a plant. A typical example of a borderline type of process operation that has pros and cons for both centralization and decentralization into cells would be injection molding.

Always keep in mind that centralized functions tend to *breed* inventories. Inventories form a relatively large portion of current assets and therefore bear a significant financial cost to a corporation. A company with centralized operations may be able to achieve some minor direct labor cost advantages over cells. However, the reductions in inventories and indirect costs, coupled with higher quality and a closer customer focus, usually make cellular arrangements a better choice. The most efficient companies have a mix of centralized departmental-type plant arrangements (for technical and capital-intensive operations) and cellular or paced-line types of layouts for flow-through small-batch operations and assembly (low skill, labor-intensive operations).

5

MATERIALS HANDLING ANALYSIS

BULK VERSUS UNIT MATERIALS HANDLING

One of the primary thrusts of this book is that of developing optimum plant layouts while minimizing *total costs* to the enterprise. As we discussed previously, there are other important factors besides materials handling costs that have an important bearing on plant layouts. For the purposes of discussing materials handling analysis and equipment, however, we will assume those other decisions have been made and manufacturing philosophies determined. We now need to concentrate on minimizing handling costs.

The layouts we address typically involve the handling of discrete "unit" loads of materials. These loads may be very tiny or relatively large. Typical examples of containers for transporting discrete loads within a manufacturing plant are shown in Figure 5-1. Usually (but not always), the handling of discrete loads involves much more labor than automatic feeds or piped flows of bulk materials. That labor may be non-exertional, such as driving a forklift truck, or it may be highly exertional, such as pushing a cart by hand. The handling of bulk materials, such as gravel, coal, bulk liquids, and sand are complex engineering subjects of their own and beyond the scope of this book. However, the handling of bulk materials that have been packaged into discrete handling containers, such as bags, drums, etc., *is* within our scope here.

Obviously, one should always work toward reducing the amount of human physical labor and exertion required to move and transport materials. This not only makes sense from a hu-

manistic point of view, but from a purely ergonomic and injury-prevention perspective as well. It is this philosophy that pervades the content of this book.

THE BASIC QUESTIONS FOR WORK SIMPLIFICATION

Before determining the materials handling systems and layouts that we will employ in our plant, it makes sense to review the existing manufacturing operations and simplify them wherever possible. In other words, the planner should not jump into developing layouts before trying to improve the current or proposed operations. There are several old clichés that every layout planner and materials handling engineer should remember and live by. One is the KISS system, which means "keep it simple and sane." Another is the adage that *the best materials handling system is no materials handling system at all.* That thought stems from the fact that materials handling, *per se*, adds no value to the part or product handled, but it does add cost. The basic questions to be asked for all activities and materials movements are:

- *What* operation are we doing and what materials are we moving? Can we combine this operation with another one and eliminate the movement of material? For example, can an operator at the first production operation replace the need for a separate, standalone, incoming inspection function? Can a machine operator perform value-added assembly work during a machine cycle in lieu of just having an idle wait period?

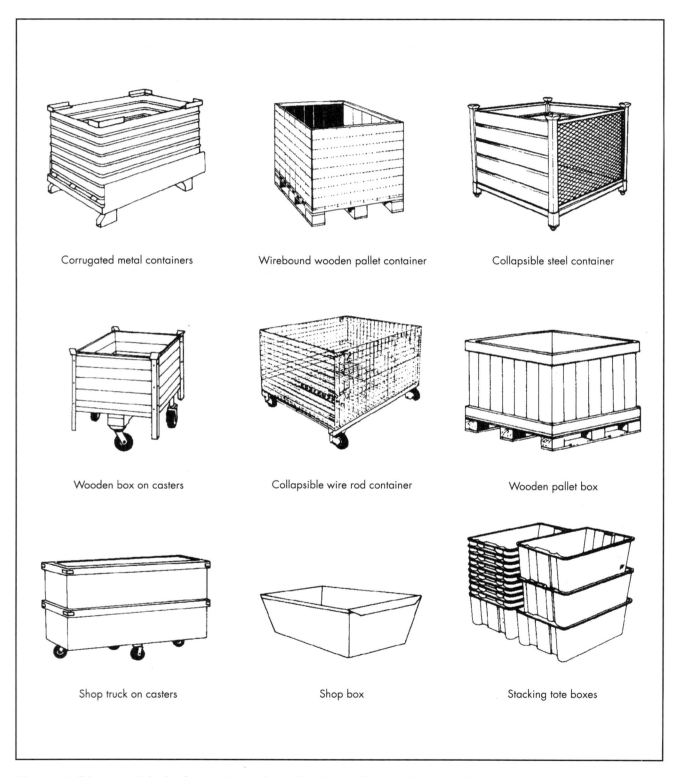

Corrugated metal containers

Wirebound wooden pallet container

Collapsible steel container

Wooden box on casters

Collapsible wire rod container

Wooden pallet box

Shop truck on casters

Shop box

Stacking tote boxes

Figure 5-1. *Discrete unit loads of materials may be small or large. Their size determines the type of containers used to transport them. Typical are the containers shown.*

- *Why* are we performing the operation and why are we moving materials at all? Can we eliminate the operation or the handling altogether? Can we simplify it? For example, can we integrate all inspection operations into the production operations? Can we use returnable containers to eliminate the dunnage materials and labor involved in unnecessary packaging or unpackaging at the shipping and receiving areas?
- *Where* do we need to move the material? Can we shorten the distance? Can we move several items at a time to reduce the number of trips? Are there straight-running aisles? If not, why not? Can we have point-of-use delivery and storage?
- *When* and how often are we performing the operation and when do we need to move the material? Can the schedule be changed to reduce the number of handlings without affecting customer service or "starving" the next operation?
- *How* are we performing the operation or moving the material? Can we redesign the handling container to help ourselves? How many nonvalue-added pickups and setdowns are we making? *Count them and reduce them!* Do we have "dead heading" return trips for materials handling equipment? If so, can we correct the situation? What is the actual time utilization of our mobile equipment?
- *Who* is performing the operation or moving the material? Do we have two-person lifting jobs? Are we using skilled labor for moving materials? Are we using direct or indirect labor for moving materials? Can the secondary operation or materials movement be done within a machine cycle? Are we losing productivity because of the time it takes to move materials? Will the vendor bring materials directly to the point of use?

It makes little sense to duplicate current problems without attempting improvements first. Frequently, time constraints do not allow a full, detailed work-simplification analysis, but that is no excuse for not trying. The development of a new layout offers the company a golden opportunity for improvements. That opportunity may not present itself again for a very long time.

Work simplification is an important prerequisite to any proposed plant rearrangement or revised layout. In addition to the *what, why, where, when, how,* and *who* questions, the four key tasks involved in work simplification are easy to remember. They can be summed up into the acronym SECS.

> **S**tudy
> **E**liminate
> **C**ombine
> **S**implify

Work simplification is discussed again when we review operations process charts later.

LARGE VERSUS SMALL UNIT LOADS

More than a quarter of a century ago, the Material Handling Institute and the U.S.-based College-Industry Committee on Material Handling Education published "The Principles of Materials Handling." The 20 original principles were essentially fundamental rules derived from the common-sense practices followed by many advanced companies and consultants of that era. The principles were first organized and "documented" by the late James M. Apple, Sr., a true pioneering educator from Georgia Institute of Technology, and have been expanded by others as necessary.* To Professor Apple's credit, most of the original 20 principles are still valid today, with the possible exception of "the unit size principle."

The original unit size principle recommended that one should always try to "increase the quantity, size, and weight of the load handled." The purpose being, of course, to reduce the total number of movements or trips required to move or transport a given quantity of materials. Theoretically, observance of that principle would serve to lower materials handling costs. If you are involved with bulk materials handling systems, or *external* physical distribution logistics (over-the-road trucking, ship cargo, etc.), that principle holds absolutely. There are exceptions, such as when working with explosives or some other hazardous materials that could cause

---

*For an up-to-date listing, contact The Material Handling Institute, Inc., Charlotte, North Carolina.

catastrophic damage. If you are working with *internal* discrete unit loads in manufacturing, the "large-load" principle may not hold. While still mathematically correct, the large-load principle can lead the *total manufacturing enterprise* into trouble if applied incorrectly. If materials handling is not analyzed as a total system, which includes both movement and storage, employing the large-load principle will *raise* overall costs, not lower them.

What was not recognized by the educators and consultants a quarter of a century ago was the deleterious effect the observance of this principle had on the *overall* internal costs incurred by the enterprise. This large-load principle was "pounded" into the heads of young manufacturing and industrial engineers for years and, unfortunately, is still being espoused by some educators. The potentially harmful downstream effects of using large batches and handling loads of discrete items is still not recognized by many manufacturers.

Why can large loads lead a manufacturer into trouble? Think for a minute about the physical effect on WIP inventories. Suppose I am making a subassembly that is 0.5 ft^3 (0.014 m^3) in size and I have 10 separate manufacturing operations to perform to complete each subassembly. Let's also suppose I am working with a right-hand and left-hand unit—the complete product consists of two mirror-image subassemblies. I can either plan one manufacturing cell with, say, a batch size of two subassemblies (1 ft^3 [0.03 m^3] in volume) or a traditional, departmentalized manufacturing arrangement with 10 departments and a batch size of 256 subassemblies (128 product pair units). I selected a 256-subassembly batch because we are following the large-load principle and can move the entire batch in one fork truck movement (carrying two 64 ft^3 [1.81 m^3] pallets at a time). We have not discussed how we are physically going to move each of the two-piece batches, but let's assume that the large-load principle clearly has much lower materials handling *transport* costs associated with it. We now have to ask some general questions about the two different manufacturing and handling systems.

- Which system will carry a higher inventory and need the most floor space to house the inventory?
- Which system is more likely to have the highest inventory carrying costs?
- If a defective assembly is found at the shipping dock, which system is liable to result in the most scrap and rework costs?
- Which system is more likely to need a pallet rack installation to store materials?
- Which system will probably need a more sophisticated material control hardware/software system?
- Which system would be easier to schedule and expedite?
- Would the small-batch mode of operations require a forklift truck?
- What are the annual depreciation or lease costs for the fork truck?
- What are the annual costs of maintaining the fork truck and training its operators?
- What does the operator of the fork truck do when he or she is not moving materials?
- Which system would probably incur the most product damage costs in handling operations?
- How many lead persons will we need with each system?

Admittedly, these questions are biased in favor of small unit loads. That bias recognized, you should agree that, unless we were consuming an inordinate amount of internal resources in moving the individual small batches, the large-load principle would not be the best choice. Clearly, this condition changes when the products reach the shipping dock. At the shipping dock, the large-load principle usually (but not always, depending on the wishes of the customer) comes back into play. Notwithstanding a just-in-time delivery schedule, either you or your customer may want to accumulate large loads or batch-mixed loads at the shipping point to gain the advantage of lower freight costs.

Every manufacturing plant will have its own special situations where the typical rules do not apply or may be overridden by other more important factors. The important point to remember is that larger unit load sizes usually lower materials handling transport costs but have a tendency to increase costs in other areas.

THE EFFECTS OF UNIT LOAD AND CONTAINER CONFIGURATION

The lot sizes used in a manufacturing cycle are frequently determined by factors beyond one's immediate control. Typical examples of controlling processes are batch chemical processes or batch heat-treating processes that have a predetermined mass quantity input.

I recently worked with an automotive machined-parts supplier which employed a batch heat-treat furnace in the manufacturing process. The company had excellent experience in its heat-treating processes and had almost a defect-free history. The heat-treat department was set up as a separate business unit within the plant. The company also had several other focused business units within the plant, one of which used the heat-treat unit's services on a daily basis. One furnace, which could heat-treat 200 specific parts in one batch, was dedicated to one of the business units that performed machining operations.

In the machining business units, prior to heat-treat, the company had listened to the latest buzz word of the day and spent an inordinate amount of money and effort to convert their machining line to single-part flow. The line had previously been set up for an optimum batch size and flow of 25 pieces. In the original system, individual basket containers, holding 25 pieces, were shuttled from one machine to another. In the revised system, single parts moved from one position to the next. The company had taken the small unit load concept to its ultimate extreme. The company did eliminate a 2-day storage buffer between the machining operations and heat-treat (they could have accomplished that without going to a single-part flow system). Also, they were able to reduce some local WIP inventories (at the cost of many more handling operations within the operation). However, no matter how small the size of the batch coming to the heat-treat unit, the parts still had to be queued up until 200 pieces could be cleaned and accumulated for the furnace batch.

The heat-treat furnace controls were then altered to allow heat-treating of smaller batches, but quality suffered. Company management returned the furnace controls to their original state to optimize the process at a mass of 200 parts. However, instead of putting 200 good production parts in the furnace, they "slugged" the furnace with scrap steel parts having an equivalent weight of 100 parts. This gave them a "good" batch size of 100 parts. This, in turn, cut the daily capacity throughput of the particular business unit by 50% and increased total heat-treat costs by 100% on a per-unit basis. When I last talked with the company, they were considering subcontracting some of their machining operations. Why? Because they could not meet the production schedule swings that their customer was imposing with a "no inventory" policy and the single-part flow concept. They were also considering closing down their heat-treating operations because they could now get the heat-treating done less expensively from a vendor. (The outside heat-treat source could not match their internal quality nor their internal costs at the original 200-piece batch level.)

Obviously, one can take any good concept and push it to extremes. In a manufacturing environment with constantly changing setups and model variations, the container size used can be very important. Figure 5-2 shows the potential effect of container size or handling increment on lot cycle times for various lot quantities. In all cases, we assume there is a transfer to the next operation *only* when the handling container is full. Let's look at the pros and cons of using small handling containers in a cell or assembly-line operation.

On the positive side of the ledger:

- Less total line length required.
- Containers may be moved without mechanical assist.
- There are fewer returns of unused parts to the stockroom, resulting in increased control.
- More staging area is available for mixed model lines.
- Faster parts changeovers.
- Supports JIT/*kanban* practices.

The drawbacks are:

- Requires higher amount of replenishment labor.
- May require purchase, storage, and maintenance of many tote-size containers.
- May reduce amount of line expansion flexibility for new products.

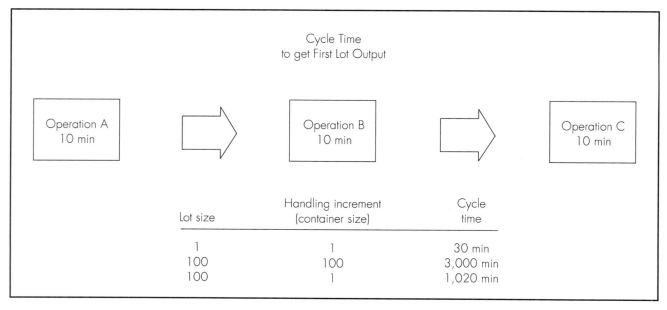

Figure 5-2. *Lot size and handling increment can have a dramatic effect on cycle time.*

When considering any kind of materials handling plan within a complex manufacturing operation, one may easily get overwhelmed with the myriad detail requirements for handling a multitude of parts or products. This problem typically arises when we try to develop a plant-wide materials handling mechanization scheme. Frequently, it helps to categorize or classify the various parts to be handled into manageable groups. It may be far better to concentrate on a common container, the design of which focuses on easily-mechanized transport. The container would be configured to normally hold as many varied parts as possible. We can call such a design a *common denominator handling unit*. In developing such a unit, certain considerations or questions arise.

- Is it possible to develop a single, standardized container or tote box configuration for most parts, products, and components? Can we design the container for conveyor transport or have it include a mechanical interface for automated lifting?
- Can we develop a universal container to handle the great majority (perhaps 80%) of the items (the 80/20 rule)?
- Is it possible to balance the production rates and product cube movements between individual work centers so that the handling system can become the in-process production bank?
- If we are using a traditional approach, our container will tend to be biased towards a large size (which may be correct in some situations). If we are considering a JIT approach, we should consider the use of the smallest-size-acceptable configuration.
- We also should investigate how we can best handle knockdown or nest containers so we can utilize cube space effectively for both loaded and unloaded containers.

An example of a stackable and nestable unit rack is shown in Figure 5-3. This rack is used to stack items that, by themselves, would normally not be stackable. The stacking allows maximum use of cube space within the operation.

DUAL MANUFACTURING AND SHIPPING CONTAINERS

A Special Case

Before leaving the subject of containers, we should add one note of caution about the design of returnable-type containers that are shipped to customers. Sometimes containers or

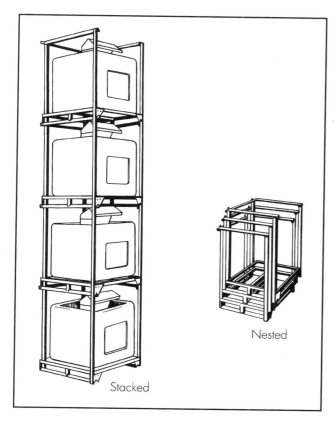

Nested

Stacked

Figure 5-3. Stackable/nestable racks make efficient use of cube space, both in use and when awaiting use.

movable cart/containers can be designed *too* well. I have seen several instances where stackable carts and containers were so well designed that customers hoarded them and used them for their own storage system. In other words, the "returnable" containers were not always returned. The customers used them in their own plants instead of installing pallet racks or other storage devices.

I recall one company in particular that was producing a very competitive commodity-type of building product (approximately 12 to 16 ft [3.65 to 4.88 m] in length). As an aid to handling the product within the plant, and in simplifying the loading of truck shipments, the company developed a very well-designed, hand- and fork-lift-truck-movable wheeled cart that was stackable as well. Instead of the previous time-consuming process of hand loading long cartons of product on trucks, they could now use a fork-

lift truck with a boom attachment to load the carts of bulk (uncartoned) product directly onto shipping trucks. The manufacturer also saved money by eliminating unnecessary cartoning materials. Likewise, at the customer's site, the carts could be quickly offloaded from the delivery truck, saving a significant amount of labor and handling of dunnage.

The customers were supposed to remove the product, knock down the carts for compact nesting, and then return them. However, they liked the stackable carts so much they did not remove the product immediately but instead *used the stackable carts for their own storage* of parts at the beginning of their fabrication processes. They would only pull a cart out of service when it was nearly empty and another full one was available to replace the empty one. When the manufacturer complained about the hoarding of carts, the customers essentially dictated that unless they were allowed to keep them, they would buy from another supplier. The stackable, wheeled carts cost more than $900 each and the manufacturer was stuck with a perpetual cart inventory at suppliers valued at more than $1 million! To add insult to injury, subsequent field visits disclosed that some of the customers were even storing product from several of the manufacturer's competitors in the carts. Again, one must be careful to avoid traps such as this when implementing a returnable container policy. At the very least, a strongly worded pricing or lease arrangement with the customers for holding returnable containers should be instituted *before* implementing such a program.

TYPICAL MANUFACTURING PLANT FLOW PATTERNS

While the specific layout depends on many factors, including the configuration or footprint of the facility chosen as the location, the general patterns of layout alternatives serve as a base for further decisions. Specific amounts of space and the shape of the space assigned to each work area depend on handling and equipment as well as other factors. Nonetheless, the flow patterns shown in Figure 5-4 are common in many industries and even more common within given departmental areas.

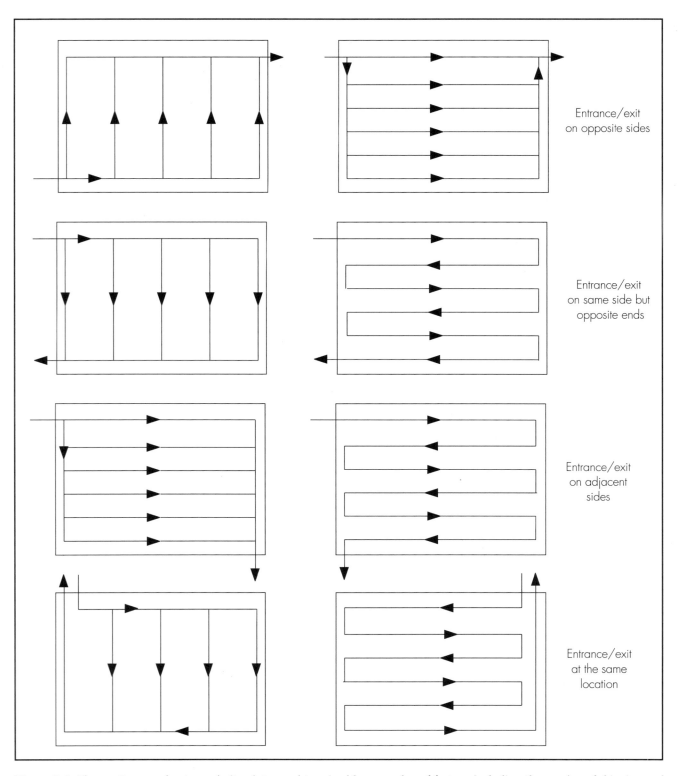

Entrance/exit on opposite sides

Entrance/exit on same side but opposite ends

Entrance/exit on adjacent sides

Entrance/exit at the same location

Figure 5-4. *Flow patterns and entrance/exit points are determined by a number of factors, including the number of shipping and receiving transactions, tradition, company philosophy, materials handling needs, and the equipment that does the handling.*

It is quite normal for small to medium-sized plants to combine shipping and receiving operations in one general area. This situation is illustrated in Figure 5-4, where the overall material flow starts and ends at the same spot on the plant layout. The reasons most cited for this "combination" pertain to the sharing of truck docks and personnel duties. There are certain industries where government regulations may prohibit the combination of receiving and shipping areas. For example, pharmaceutical companies working with ethical drugs are usually prohibited from combining these operations.

Almost by definition, combined shipping and receiving operations tend to result in an overall U-shaped flow within the plant. Sometimes this results in more two-way traffic within the plant, but as seen in Figure 5-4, that may be an oversimplification.

Most larger manufacturing plants tend to separate the receiving and shipping functions. This split allows for better material control and improved security and takes advantage of the opportunity to specialize duties and labor. The great majority of large plants direct most of their materials to be delivered to a central receiving point and finished products are shipped from a central shipping point. This centralization of activities is usually done for materials control purposes. However, some large plants have tried moving away from centralized receipt functions. Though multiple receipt points for large plants offer many benefits, this approach does require very disciplined purchasing, materials control, and quality systems.

The potential flow pattern *not* shown in Figure 5-4 involves multiple receipt and shipping doors. In some very large assembly plants, for example those of automobile manufacturers, receiving doors may be spread across an entire building wall of more than 1,500 ft (450 m). In those cases, the plant usually receives materials from trusted (and proven) *certified* suppliers on a JIT basis, delivered close to the point of usage on the assembly line. Typically, only minimum inventories (a few hours' supply) are stocked at the various receiving points, thereby freeing up a tremendous amount of formerly used storage space within the plant. Needless to emphasize, this type of plan also provides a significant reduction in inventory carrying costs and asset financing. However, the multiple-receipt system does place a heavier control and security burden on the materials management functions. It also places extreme pressure on the suppliers. Any major quality defect or slip in the shipment schedule has the potential to shut the entire plant down. Nevertheless, it has the potential for working very well in large plants. The key elements in this approach are *schedule, quality,* and the use of *trusted and certified suppliers.*

Figure 5-5 shows a sampling of some of the various materials flow patterns that have been used in manufacturing plants. Large plants generally have long, straight-through aisles and materials flow patterns with offshoots to either side. This is frequently referred to as the "spine" approach with shipping and receiving at opposite ends of the plant. Plants also may use a spine approach after completing processes fed by multiple material receipt locations.

THE OPERATIONS PROCESS (FLOW) CHART

The operations process chart (OPC) is arguably the most useful tool in materials handling analysis, work simplification, and the development of plant layouts. OPC symbol convention extractions from The American Society of Mechanical Engineers (ASME) standards are included in Figure 5-6. The figure shows some of the conventions used in making process charts. An example of an operations process chart that tracks weight is shown in Figure 5-7.

In industrial layouts, process becomes a major consideration. The method of organizing this process data into a useful format differs with the amount of data to be organized. Often, an operations process chart similar to that shown in Figure 5-8 is used when the number of products is limited or if only one particular department is being looked at, such as an assembly department. As the number of products to be manufactured increases, data is frequently organized on a multiproduct process chart. An example multiproduct operations process chart is shown in Figure 5-9. Sometimes a selected sample of products or representative items is used in the chart, rather than the entire catalog of products produced.

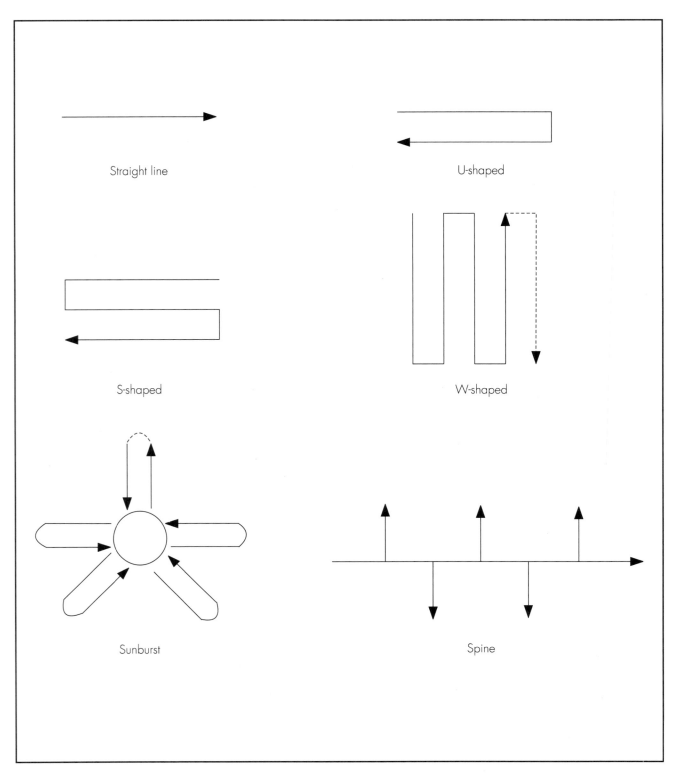

Figure 5-5. *The geometry of materials flow in manufacturing traverses the spectrum of pattern and shape, dictated largely by plant size.*

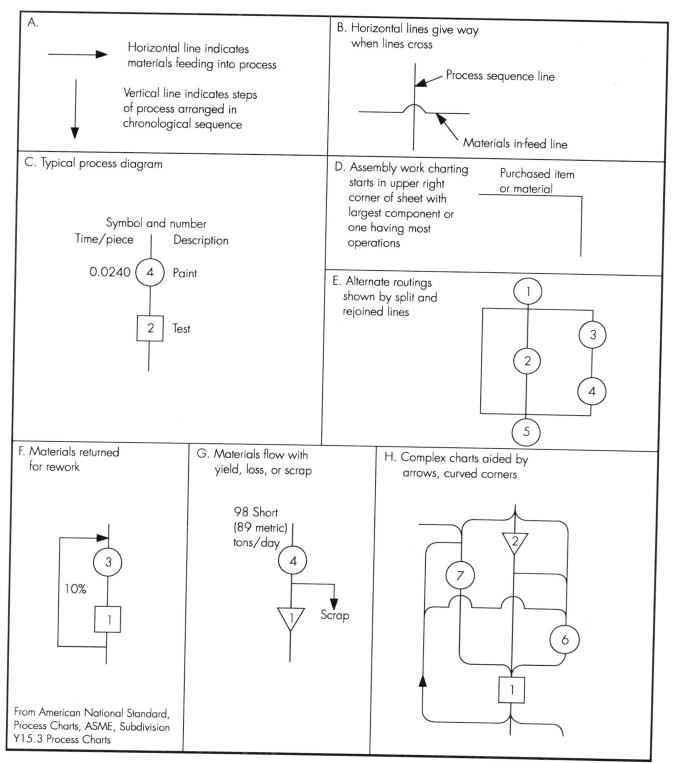

A.

Horizontal line indicates materials feeding into process

Vertical line indicates steps of process arranged in chronological sequence

B. Horizontal lines give way when lines cross

Process sequence line

Materials in-feed line

C. Typical process diagram

Symbol and number

Time/piece   Description

0.0240   4   Paint

2   Test

D. Assembly work charting starts in upper right corner of sheet with largest component or one having most operations

Purchased item or material

E. Alternate routings shown by split and rejoined lines

F. Materials returned for rework

10%

From American National Standard, Process Charts, ASME, Subdivision Y15.3 Process Charts

G. Materials flow with yield, loss, or scrap

98 Short (89 metric) tons/day

Scrap

H. Complex charts aided by arrows, curved corners

Figure 5-6. *Simplifying (and standardizing) materials handling analysis, operation process chart conventions help form the basis for a relational layout. (Courtesy ASME)*

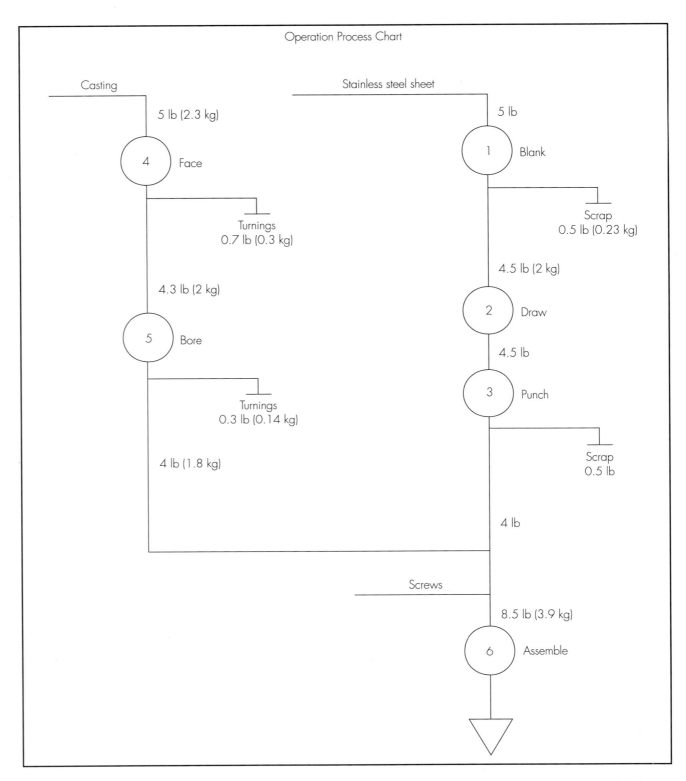

Figure 5-7. *Operations process charts can be used to track and map specific measurables, such as this one tracking weights through manufacturing operations.*

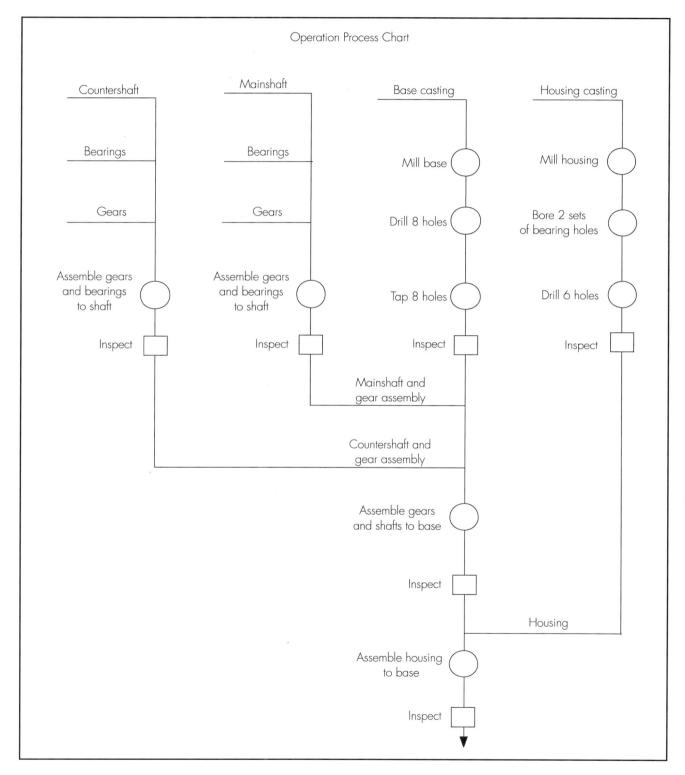

Figure 5-8. *Specialized operation process charts can be used when a company's product offering is limited or when a particular department is highlighted, such as the assembly operation shown here.*

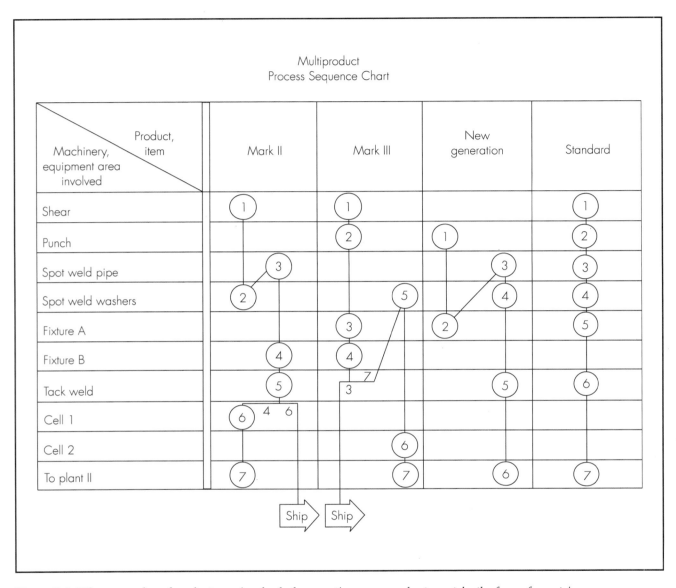

Figure 5-9. *When a number of products are involved, the operations process chart can take the form of a matrix.*

When a large number of products or parts make using a multiproduct process chart impractical, the items produced are sometimes grouped by families or by items requiring similar processes. These familiarities or similar process groups become the basis for a *from-to* chart (discussed later).

On very large projects, a consolidated operations process chart is typically developed to show the major material flows about the site or about the plant. It is recommended that these show major material flowpaths only in the initial project stages. A typical example, taken from a large shipyard project, is shown in Figure 5-10. A more complex example from another large shipyard is shown in Figure 5-11. In the complex case, many minor flowpaths as well as major material flowpaths are shown.

Occasionally, the layout planner may want to chart a selection of worst-condition items (rather than all items) to ensure that all products can be produced within the proposed layout. Such a

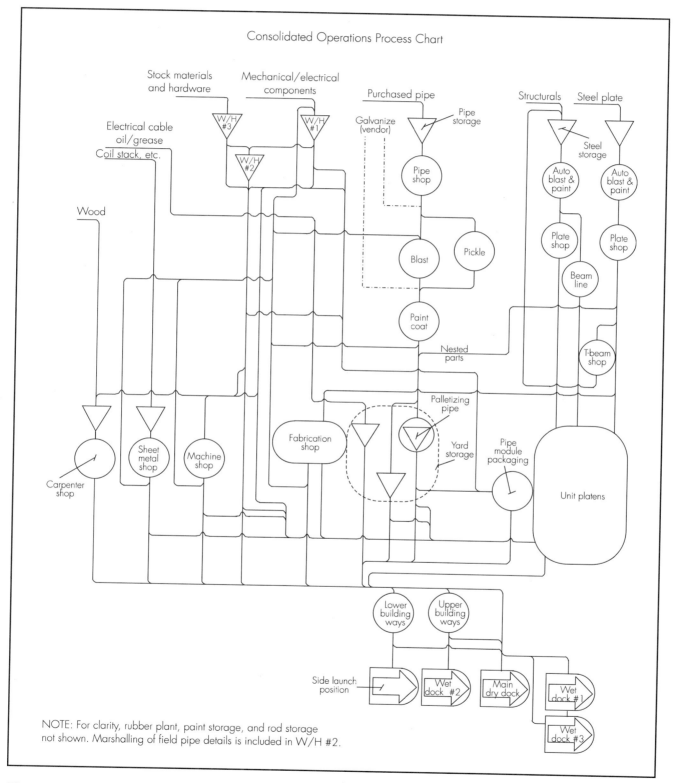

Figure 5-10. *Consolidated operations process charts show major materials flow on the site or inside the plant for very large projects.*

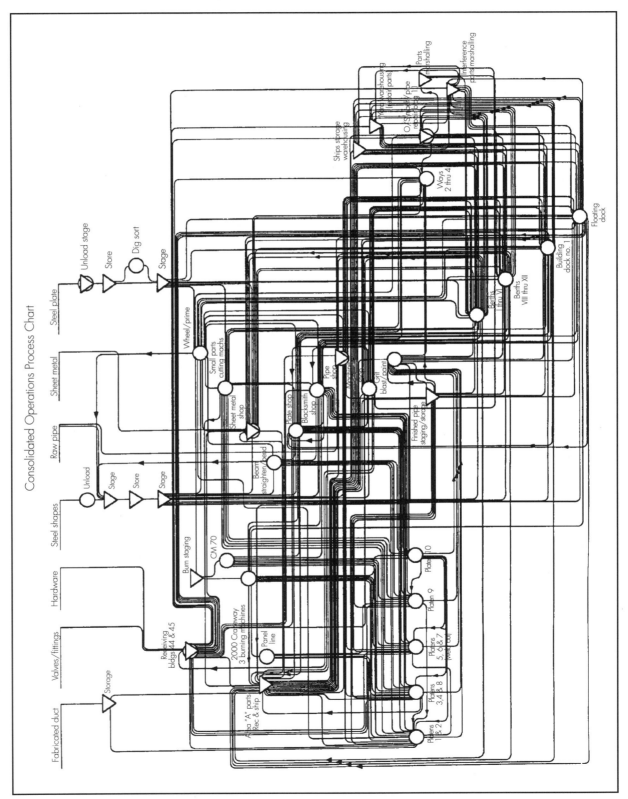

Figure 5-11. *Consolidated operations process charts such as this one which includes major as well as many minor flowpaths, though comprehensive, are sometimes difficult to read. (Courtesy National Steel and Shipbuilding Co.)*

selection can include three to five items, ranked worst or highest in each or several of the following categories.

(a) Heaviest

(b) Bulkiest

(c) Most fragile

(d) Value

(e) Greatest volume

(f) Largest number of operations required

(g) Most dangerous

(h) Most difficult to handle

(i) Worst quality assurance problems

(j) Worst scrap and reject record

Frequently, a graphic representation of the materials flow can be drawn or sketched directly on the operations process chart. In some smaller factories, the operations process chart may become the diagram for the *actual* plant layout or at least the major material flow patterns within the plant. This is not recommended in all cases, since serious blunders may occur if the relative strengths of the flows are ignored and only straight-line connecting paths are considered.

It is important to recognize that materials flow can be represented on the operations process chart on both a sequential basis and a magnitude basis. Since the magnitude of materials flow helps form the degree of "closeness" (relationships) between activities, the operations process chart is usually a necessity for developing optimum plant layouts. It is normally the precursor to the development of a from-to chart (a highly useful tool for organizing data, discussed in detail later in this chapter).

Frequently, traditional OPCs do not show *all* potential materials flows. The planner must remember to take into account those important flows that would normally not appear in the mainstream of activity but could, in fact, cause problems later on. Scrap from press shops, cardboard from unpacking operations, empty containers or pallets, etc., should all be checked out thoroughly for handling volume and methods. A flow that is quite frequently overlooked is the internal return flow of empty containers (e.g., totes, baskets, pallets, etc.) to the starting point of the operation. I have encountered dozens of situations in which starting processes in a plant are "starved" for handling containers and have to stop operations because someone neglected to bring empty containers back to the starting

point in the operation. A simple oversight like this invariably results in lost capacity or lost production that can never be regained. Even with automated container return systems, it seems to be a natural fact of life that such "starvation" occurs—it just seems natural for people to accumulate and hoard empty handling containers. This is a management and discipline problem that many plant managers frequently face. In any case, it is arguably as important that the flowpath of supporting materials such as empty containers be noted on operations process charts as the flow of direct materials themselves.

Similarly, in many plants, the flow of discarded packaging materials can be more of a handling problem than the flow of the actual materials that arrived in the packages. This is a common problem in some industries, particularly when the detrashing process does not occur upon receipt at the receiving dock. These operations and flows are frequently overlooked on most operations process charts. Likewise, nonmainstream operations that may occur are frequently missed, such as a return flow or rework loop after an inspection operation. Since rejected parts are not commonly anticipated, these types of flowpaths may be easily overlooked.

It is important to bear in mind that the OPC serves many valuable functions. As you, the planner, investigate the process steps, you will become intimately familiar with the overall shop-floor processes and any current or proposed materials handling methods. You also should introduce work simplification methods at this stage if you have not done so already. Finally, the OPC will highlight the material flowpaths. As we will see later, the intensity of the materials flow along these paths forms a major component of the affinities or closeness required between activities.

THE IMPORTANCE AND OPPORTUNITIES OF PRODUCTION LINE AND CELL BALANCES

Although we have discussed consolidated operations process charts for entire plants and entire sites, it is also important to know about detailed operations process charts or sequential routing charts, as they are frequently called.

Unfortunately, sequential OPCs or routing charts for a particular manufacturing cell or assembly operation are often misused. The planner should refrain from taking a chart like the one shown in Figure 5-8 as a given. He or she should first question the process. By "question," I mean that the principles of work simplification should be applied. The basic questions, outlined in the beginning of this chapter, should be addressed. It may be a mistake (or lost opportunity) to take the OPC at face value and lay out the detailed work centers in the same sequential relationship as depicted on the OPC. Let's take a simple example.

The numbers noted beside the bubbles shown in Figure 5-12 indicate the standard times allowed for those operations. Whether we use predetermined, engineered time standards for new operations or actual measured time from stopwatch studies is not important. If the objective is to design an efficient operation, it is important that the standard times be known and that they be accurate. If we take Figure 5-12 at face value, we might simply combine the operations as shown in Figure 5-13 to get a 9-minute cycle. The third set of combined operations would have 3 minutes of slack time or idle time but the first two sets of combined operations would have zero idle time. The *total production and total capacity of the operation would be capped at 160 pieces per day*. The calculations at the bottom of Figure 5-13 show the results of the combination using three operators on each of three work shifts. But is that the best we can do?

If instead we change the proposed layout and combine the operations to achieve a 6-minute cycle, we can potentially achieve much better productivity. In fact, with only a minor change in the layout, we can *totally eliminate the third-shift portion of this operation*. We achieve the same overall current production, with less overall people, on two shifts. With this seemingly minor change, *we also increase total capacity by 50%, to 240 pieces per day*. How can we physically achieve that magnitude of change without speeding up the work pace and alienating the employees? Figure 5-14 shows the new combined operations. The operations are now performed with four people on each of two shifts

(eight people total compared to nine previously). The savings results are tabulated in Figure 5-15.

It's important to grasp the full importance of this simple example. Keep in mind that we did not change the detailed operations themselves. The standard hours or minutes of work content remained the same. All we changed were (a) the individual operator responsibilities within the cell and (b) the layout to facilitate the new combinations. These kinds of dramatic potential improvements are often overlooked when the analysis is left up to the work teams within a cell, particularly when standard times have not been published. Many people fighting day-to-day production problems cannot think "beyond the box," so to speak. In their minds, adding people to shifts to increase productivity just does not make sense.

It does not help that one of the 14 principles espoused by Deming was the elimination of time standards. That principle may apply very well in some cases, but it does not apply when you are trying to analyze and improve a particular situation. To be successful, you need a baseline measurement from which to gage progress or change.

In trying to achieve optimum labor balances, many planners attempt to avoid the use of buffer storage within cells or on production lines. In actual practice, temporary buffers can be very beneficial. There is an obvious benefit if a piece of equipment breaks down. The workers in positions prior to the breakdown can continue to build to a buffer storage. The workers at positions past the breakdown can continue to work from the previous buffer. This also emphasizes the point that each successive operation on a progressive line should have a slightly higher capacity than the preceding operation, if at all possible. Obviously, this analysis assumes that you need to build the product based on demand, not on unnecessary inventories.

Buffer storage can help in other, more subtle, ways as well. If there is a rigid assignment of responsibilities and the labor balance is set to accommodate the pace of an "average" worker, variances in operator performance are sure to occur. A replacement operator may perform only at 90% of the "average" of the operator who was replaced. Inexperienced, temporary workers

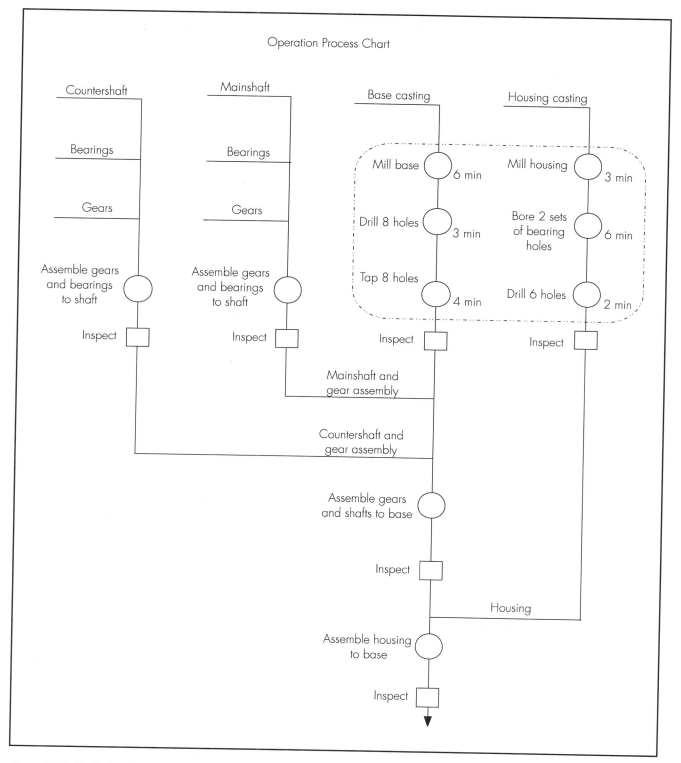

Figure 5-12. *To design the most efficient operation, it is crucial to know the standard times for completion of specific operations, as shown here within the broken lines.*

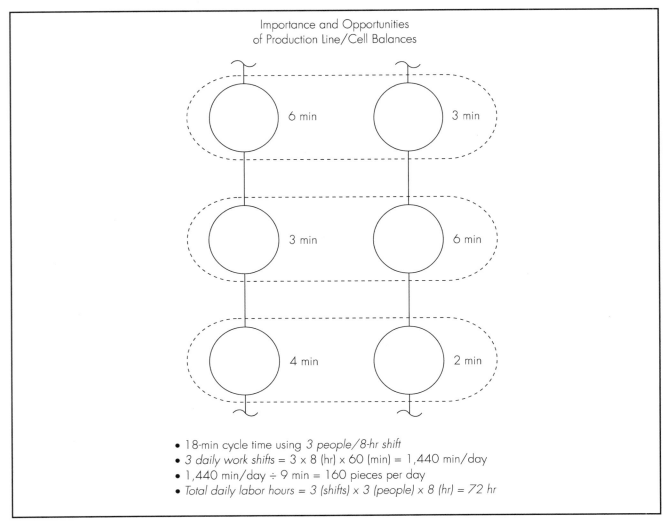

Importance and Opportunities
of Production Line/Cell Balances

6 min   3 min

3 min   6 min

4 min   2 min

- 18-min cycle time using *3 people/8-hr shift*
- *3 daily work shifts* = 3 x 8 (hr) x 60 (min) = 1,440 min/day
- 1,440 min/day ÷ 9 min = 160 pieces per day
- *Total daily labor hours* = 3 (shifts) x 3 (people) x 8 (hr) = *72 hr*

Figure 5-13. *By combining operations (as indicated within the broken lines) the planner is able to evaluate the labor balance and current capacity in terms of where improvements can be made.*

may perform at only a 55 to 65% level of a "normal" labor standard (the standard being a "fair day's work"). In fact, if worker performance levels can be assumed to follow a normal distribution, then 50% of the operators will have performance levels below the average (see Figure 5-16). This is a major problem with sequence-dependent lines or cells.

Assuming no buffers are allowed and the slowest operator is placed at the last or end position on the line, the entire line or cell will be forced to slow down to the pace of the slow operator. If the slowest individual is placed at the start of the line, the individuals after that point will be starved for work and the line output will still be limited to the output of the slowest operator. However, let's assume the operator with the highest performance level is placed at the starting position on the line with the slowest operator at the output end of the line. If we employ a U-shaped layout arrangement and temporary buffers are allowed, enough of a buffer may be built up at each station to allow the fast operator to leave his or her position and help the slow operator (presumably, in the U configuration they are positioned close to one

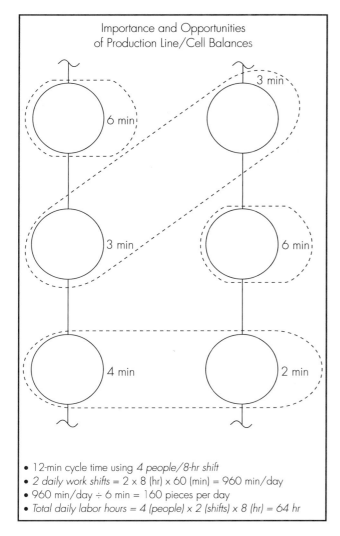

Importance and Opportunities
of Production Line/Cell Balances

- 12-min cycle time using *4 people/8-hr shift*
- *2 daily work shifts* = 2 x 8 (hr) x 60 (min) = 960 min/day
- *960 min/day ÷ 6 min = 160 pieces per day*
- *Total daily labor hours = 4 (people) x 2 (shifts) x 8 (hr) = 64 hr*

Figure 5-14. *By altering the layout to achieve a 6-minute cycle through a different combination of operations, significant gains in productivity can be achieved.*

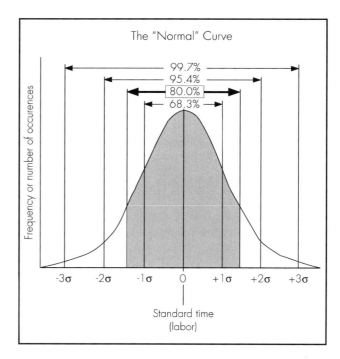

Importance and Opportunities
of Production Line/Cell Balances

For the same production output:

- Labor reduction (hours) = $\dfrac{72 - 64}{72} \times 100 = 11\%$

- Cycle time improvement = $\dfrac{9 - 6}{9} \times 100 = 33\%$

- Reduced headcount, fringes, payroll taxes $= \dfrac{9 - 8}{9} \times 100 = 11\%$

- Support services associated with a third shift eliminated

- Total capacity is doubled

Figure 5-15. *A 50% increase in capacity was achieved, along with significant reductions in other cost components, through recombining operations.*

The "Normal" Curve

99.7%
95.4%
80.0%
68.3%

-3σ   -2σ   -1σ   0   +1σ   +2σ   +3σ

Frequency or number of occurences

Standard time
(labor)

Figure 5-16. *Sequence-dependent production lines or cells suffer from "normal" distribution, in which 50% of the operators may have below-average performance levels.*

another). Overall production levels will be increased, provided we allow for some level of buffer storage.

One point that is common to most well-designed cells or sequence-dependent production lines is that the choke-point or bottleneck position is placed as close as possible to the *end-point* of the process and buffers are designed into the process. If at all possible, the choke-point should never be placed at the beginning stages of the cell or line. If it is, the line or cell will never achieve its planned capacity on a continuing basis (based on both the mathematical theory of constraints and real-world experience). Likewise, in a perfect production world, the ideal line or cell would have each successive operation capable of producing at a higher capacity than the previous position.*

We return to this subject when dynamic computer simulation and equipment utilization assumptions for calculating space needs are discussed.

Numerous other mechanical attempts have been made to remove the dependency inherent in production lines and cells. Probably the most famous (and lengthy) experiment was Volvo's attempt in the late 1980s and early 1990s to assemble automobiles in place *without* the aid of a paced assembly line. The "humanistic" system, originally conceived in 1985 to eliminate worker stress associated with paced assembly lines, took several years to develop. The system was implemented in their Uddevalla plant in Sweden, with its first operating year in 1989. The Volvo method required expensive tilting fixtures for each team assembly position. By 1991, the plant was producing only about 90 autos per day. The teams were averaging approximately 32 assembly hours per vehicle. Teams of 10 workers each required an average of 8 hours to assemble 2.5 vehicles per team per day (some teams achieved four per day). Although that level of production compared favorably with other Scandinavian auto assembly plants, Volvo's production output would never come close to matching Japanese and U.S. auto manufacturers. In 1991, the Japanese and Americans were averaging 21 and 25 assembly hours, respectively, with their assembly lines.

In fact, the Volvo team experience was much more of an experiment in humanistics than in maximizing production output. However, a great number of beneficial improvements were made as a result of Volvo's work. The planning and implementation effort should be a required study for those with a keen interest in worker ergonomics and a stress-free working environment.

Volvo claimed that, under ideal conditions, they could achieve more than a 40% productivity improvement over the nonbuffered paced lines typically used in the U.S. However, Volvo was never able to achieve that improvement in actual practice. Experts who studied the experiment claimed the real reasons for not meeting production goals were the extremely high employee training requirements, coupled with the extraordinary absentee rates experienced throughout Scandinavia (when compared to absenteeism rates in Japan and the U.S.)

The Japanese automakers were very interested in Volvo's Uddevalla experience. Some Japanese authorities believe the original "lean" production system, pioneered by Toyota, is still somewhat problematic and stressful. But most agree that it is a great improvement over the traditional, regimented, paced assembly lines pioneered by Ford. The lean system was seen to be a tremendous improvement because it fostered more employee involvement.

In 1995, Toyota was producing RAV4 (sport utility) vehicles in Toyota City, Japan, using an overhead conveyor line for car bodies with several buffer areas. The inclusion of buffers was much to the chagrin of the management pundits and gurus in the U.S., who were continuing to tout single-piece flow with zero buffers. As a result of that chagrin, and with millions of dollars in U.S. consulting revenues at stake, the Japanese automaker's return to buffers was almost totally ignored by the trade press.

The new Toyota line was split into five parts with the buffer zones between. Vehicles only come into each area when the workers are ready for them (sections of the line may be shut down *without* shutting down the entire assembly line). Workers stand on moving conveyor belts that follow the car through their respective areas. Other than moving engines and transmissions to the line, there is very little automation. Although the production rate is not extremely high by U.S. standards, (estimated at 9,000 to 10,000 vehicles per month), the labor hours are said

---

*For an excellent discussion of this subject along with a layman's discussion of the theory of constraints and why typical "balanced" models do not work, I highly recommend the book: *The Goal*, by Goldratt and Cox, 2nd Revised Edition, North River Press, Inc., Croton-on-Hudson, NY, 1992.

to be less than half that of a typical U.S. assembly plant.

In the mid-1990s, Toyota, which was once heavily automated, was moving steadily away from automation. In prior automated plants, the increase in maintenance costs was much higher than anticipated. (Significant productivity improvements have come about by using common parts on many vehicles—similar to what Chrysler and Ford have been doing for many years.)

For light manufacturing and assembly, there are several methods for reducing the amount of dependency on prior operations. All of these allow operators or very small teams of operators to work at their individual or small-team performance levels. Figure 5-17 depicts an approach that uses a recirculating conveyor. If a part operation is missed, it can be "recycled" to another open position or returned to a specific location.

EQUIVALENT UNIT LOAD ANALYSIS

Richard Muther, an internationally recognized industrial engineer and author, originally developed the *Systematic Layout Planning* (SLP) method, upon which much of this book is based (I had the honor of working with Dick Muther in his consulting practice throughout the early 1980s). Muther truly can be considered a pioneer in the development of the rationale used for planning industrial facilities.

Since Muther's pioneering work in plant layout in the 1960s and 1970s, there has been confusion among practitioners as to the definition and quantification of material flow "intensity." Almost all computer-based (as well as manual) plant layout routines require some measure of intensity of material flow as an input. The definition and measurement problem arises when material flow is not homogeneous.

It is extremely difficult to establish a common measure for diverse material characteristics and diverse materials handling methods. To overcome the problem of flow and handling diversity, Muther, Haganaes, and others originally established the "mag count" method of determining equivalency. Mag count is a system of cubic volume measure of materials adjusted for other influencing factors, such as bulkiness, risk

of damage, etc. It was an attempt to establish a quantitative measure of *volume and intensity* of material flow *without regard to the material handling equipment used to transport the flow*. Although the mag count system was never universally accepted as a quantitative measure, the work and thinking that went into its development are still very important and impressive.

With the advent of high-powered personal computers, we can now work almost directly with detailed materials handling costs. These costs should take into account the acquisition and operating cost of the materials handling equipment used to transport the material. With modern computer-based techniques, these costs can be calculated relatively quickly (provided you have a first-cut layout from which to begin). Material flow measurements based on systems such as the mag count can be used only for rough-cut block layouts, since they tend to focus on flow volume only and (at least initially) ignore materials handling equipment and operating costs. But pure volume and transactions types of flow measurements (such as mag count) are still useful for designing new plants and visualizing materials flows, prior to the determination of the materials handling equipment to be used.

If you do not have access to the special (and expensive) programs for calculating materials handling costs, you still need some method to quantify materials flows and get started on planning the plant layout. Even the computer routines that calculate cost require that we make educated and experienced guesses on the potential materials handling equipment to be used between all work centers.

In establishing the relative cost of materials flow and handling in our initial macro layouts, it is useful to first establish an equivalent unit load. For example, moving a 17-ft (5-m) length of 12-in. (305-mm)-wide, 1-in. (25-mm)-thick plate of steel in a shipyard is not the same as moving a cubic foot (ft³, 0.028 m³) of lead the equivalent distance. Although they weigh approximately 700 lb (318 kg) each, their handling characteristics are entirely different. Let's assume we plan to use a fork truck to move each of these items in the same factory five times each work shift. Let's also assume we need to move each

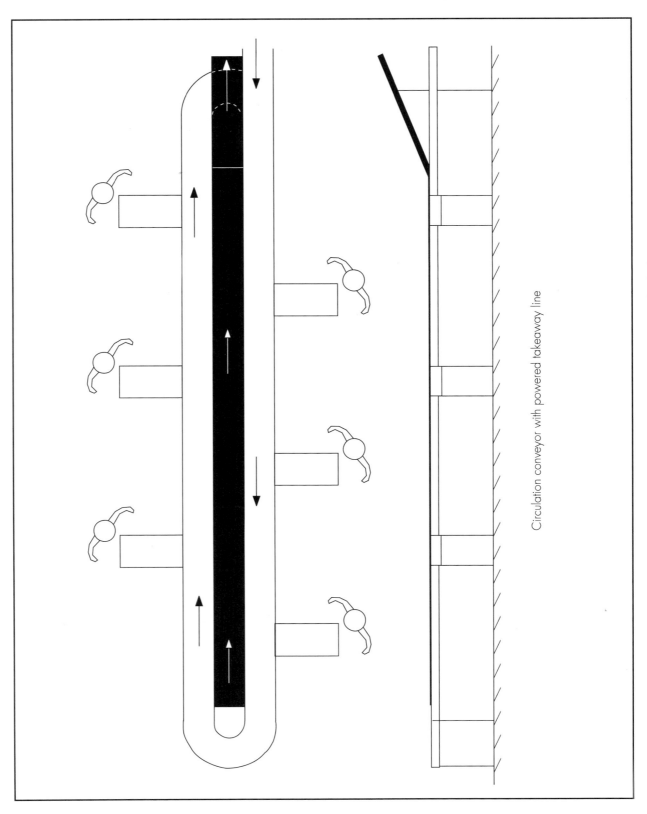

Circulation conveyor with powered takeaway line

Figure 5-17. In light manufacturing, recirculating lines enable operators or teams to work at their individual paces.

of the items between work center *A* and work center *B*. The costs to move each will be entirely different. Even though the moves include the same number of trips, the same transport distance, and the same load weight, the steel plate is a much more awkward load. It will probably take three or four times the amount of transport time compared to moving the cubic foot of lead. That time can be directly converted to higher materials handling costs for moving the plate. In the upcoming section on from-to charts, we need to assure ourselves that when we count *handling trips*, we may need to make an adjustment for the "difficulty" factor associated with the trip. In this particular case, with a difficulty factor of, say three, we would count *each single trip of the plate as equivalent to three trips of the cubic foot of lead*. The real key here is cost. If we knew the actual materials handling costs, that would automatically take care of the equivalency factor.

Similarly, in a light manufacturing environment and in lieu of having materials handling costs, some planners initially use the cubic volume to be transported as the equivalent unit load, at least to get started on a macro layout. Obviously, estimated handling cost is a better measure, since cubic volume may or may not have a direct relationship to materials handling cost.

When the materials being moved and the handling methods within a plant are diverse, the equivalent unit load concept has to be taken into account when you are nearing the detail layout stage. Some of the characteristics that may need to be considered in establishing equivalency are:

Shape	—	Awkwardness, compactness, square, flat, round, irregular
Weight	—	Per unit or specific weight
Size	—	Length, width, height
Value	—	Wood, gold
Fragility	—	Risk of damage, hazardous
Conditions	—	Sticky, wet, hot, frozen, etc.
Equipment	—	Fork truck, cart, crane, etc.

These are all taken into account when measuring handling ease, transportability, and materials handling costs. If possible, a standardized tote, pallet, or container may be used for handling materials. In that case, the focus of the layout planner shifts to the movement of "equivalent" containers (see Figure 5-1).

In many factories where forklift trucks are used extensively for materials handling, the equivalent unit load may be defined as an equivalent *pallet* load. All measured or calculated materials movements are converted into equivalent pallet (unit) loads for tracking flows. An equivalent pallet load translates into one fork truck movement. This could be the case even if the size of the load was much smaller than a regular full pallet-sized load.

Let's look at an example. In our manufacturing process, suppose we plan to use a fork truck to transport one box of parts 1.6 × 1.6 × 3.3 ft (0.5 × 0.5 × 1.0 m) in size from work center *E* to work center *F* every half hour. We realize this may be an inefficient movement and an inefficient use of the fork truck. However, due to the timing of the process and the forklift truck returning from a previous trip empty, we decide to more fully utilize the truck by picking up this load. When we initially track intensity of flow in the next section, this fork truck movement will be counted as one equivalent pallet load even though the load is relatively small. If we take an annual snapshot, we would calculate how many boxes needed to be moved each year and these would be the number of equivalent pallet loads to be handled.

Keep in mind that if we change the materials handling process, the number of equivalent pallet loads moving also may change. For instance, suppose we change our plans and accumulate two boxes of parts at work center *E* before a pickup. Let's also assume the fork truck now picks up two boxes of parts each hour instead of one. Even though the total volume of flow (the mag count) has not changed, the number of equivalent pallet load moves has been halved. Similarly, the total transport costs have theoretically been reduced.

This example clearly illustrates why one cannot divorce materials flow volumes from materials handling methods. If we manually transport a small load between work centers every 5 minutes, we would need to design the plant to have those work centers very close to each other. But if we consolidate the materials and only move between the work centers once every work shift, the work centers could be much farther apart, without increasing total materials handling costs.

In measuring materials flows on an operations process chart, the planner typically has few choices. If he or she is developing plans for a new plant, and there is no "existing" layout, the planner must base the first-cut macro analysis on the intensity of transactions or cube movements between activities (modified for known usage of materials handling equipment). Materials handling costs cannot be obtained at this early stage because we do not know the transport distances between activities until later in the plant layout process.

If the planner is fortunate enough to have an existing scaled layout or a first-pass macro layout in CAD format, actual materials handling costs can be calculated and compared for various layout alternatives. These calculations require special computer-based programs.

It is important to note that the units to be tracked in the from-to charts (discussed next) will be equivalent unit loads, wherever possible. This requires the planner to make initial assumptions on the materials handling equipment to be used.

THE FROM-TO CHART

The most encyclopedic format used to organize process data for either single or multiple products is the *from-to chart*. Such charts can indicate both volume and process flow for each item manufactured. If we are dealing with individual product data, we normally track the flows required to produce one unit. This is later extended against a production forecast to get time-based movements. If we are combining several charts, all data should be compiled for a consistent unit of time. Unless there are highly seasonal variations, most planners select an annual basis period.

The from-to chart is constructed similar to a mileage chart in this order:

1. Review the operations process chart(s) to determine those activities between which there is material flow.
2. List the activities in identical order across the top columns and down the rows on the left-hand side of the chart.

3. Establish a measure of flow that indicates equivalent unit loads or equivalent transport-related materials handling costs. When all items are equally easy to move, the projected number of trips can serve as the measure. If items vary significantly in size, shape, weight, damage potential, etc., develop transport difficulty factors to establish equivalency. If we are working on a total plant rearrangement, a sampling of moves along the paths to verify the data is recommended.
4. Using the flowpaths shown on the operations process chart(s), record the equivalent unit load moves (combine both to and from moves in one cell) between each activity pair.
5. In a multiproduct manufacturing environment, construct a "sub" from-to chart for each product (or sampling of major products). When completed, combine all of these into one "total" from-to chart.

If at all possible, it is best to list the process steps or operations in their "natural" directional sequence. The best flow would be a straight line moving from one operation to another, top to bottom. For example, a typical sequence might be:

1. Receiving
2. Staging for inspection
3. Inspection
4. Point of use storage/operation No. 1
5. Point of use storage/operation No. 2
6. Point of use storage/operation No. 3
7. Manufacturing cell No. 4
8. Manufacturing cell No. 5
9. Subassembly storage buffer
10. Cleaning
11. Painting
12. Final assembly
13. Test
14. Stage for shipping.

These functions, along with all of the supporting activities, consist of all the areas to be included in the plant layout.

It is highly recommended that from-to charts be prepared using standard computer-based spreadsheets. Normally, equivalent unit loads moving between work centers (actual or projected) are noted in intersecting cells (see Figure 5-18).

From-to Chart for All Hammers (Trips)

From \ To	Saw/stud weld	Hammer	Trim/hot punch	Trim scrap	Heat treat	Shot blast	Restrike	Restrike, salvage	Cold coin	Cold punch	Cold coin salvage	Machine	Mag and grind	O.S. machine	O.S. heat treat	O.S. other services	Store/ship	Total "from" tasks
Saw/stud weld		22,415																—
Hammer			22,415															22,415
Trim/hot punch				9,793	6,892	4,934				7		165			623		1	22,415
Trim scrap																		—
Heat treat						8,551			3				11		10			8,575
Shot blast					100		2,202	595	4,281	584	841		2,418		55	14	6,003	17,093
Restrike					1,030	10			10				1,117				210	2,377
Restrike, salvage													402				193	595
Cold coin					295	169				10			423		13		4,437	5,347
Cold punch					10				675								25	710
Cold coin salvage					245					25							816	1,086
Machine													7		158			165
Mag and grind					3	2,576	175		378	84	245				7	214	708	4,390
O.S. machine																	3	3
O.S. heat treat						853							12					865
O.S. other services														3			225	228
Store/ship																		—
Total "to" tasks	—	22,415	22,415	9,793	8,575	17,093	2,377	595	5,347	710	1,086	165	4,390	3	866	228	12,621	—

Figure 5-18. The comprehensive from-to chart can indicate both volume and process flow for each item manufactured, and represents an excellent format for organizing process data.

If you are doing a rough-cut, macro-plant layout prior to determining materials handling equipment (MHE), you may have to chart the cubic volume of materials moving first. Based on that analysis, and after developing several rough-cut macro-block layout alternatives, you should then make preliminary assumptions of the MHE to be used. A new (or several) from-to chart(s) should then be developed to show equivalent unit loads moving between work centers. The computer-based spreadsheet format is helpful in making summations both vertically and horizontally. The planner should double check these summary figures to ensure that the same quantity of materials that moves into a work center moves out of the work center (after accounting for physical changes, scrap fall off, etc.).

The *simplest* from-to chart uses a simple hash mark count for tabulating each item moving from one activity to another (see Figure 5-19). This format provides a rough volume and routing picture, but normally does not indicate the cost of materials flow for any specific product—with one exception: when a homogeneous materials handling method is used for transporting materials between all work centers. For example, assume we are planning a series of manufacturing operations and are using a simple push-cart to manually transport materials between each operation. Our from-to chart would only need to show the number of cart movements made in a given time period (or the number of movements required to complete a particular product) between work centers. Whether there is only one particular item on the cart or a dozen of another item, we are interested at this point only in the number of total cart movements and, by association, the labor costs involved in moving the carts. This simple from-to chart assumes a constant cost, per foot (meter) of distance moved, for transporting the carts. (That constant cost assumption is not *totally* true, but it is close enough for our purposes. Actually, pickup and setdown costs do not change: only the variable cost associated with transport distance changes as we alter the plant layout. Materials handling cost changes are therefore not exactly a direct linear function of distance changes.)

Keep in mind that the logic of our initial layouts, or macro layouts as they are sometimes called, will be based on minimizing materials handling costs. In other words, those work-center pairs that have a high intensity of materials flow between them (and have an associated high materials handling cost) need to be close to one another in our macro layout. Those pairs that have a low intensity of materials flow (and by association, low materials handling cost) between them can be placed farther apart in our layouts.

It is important to recognize both the required *sequence* and *intensity* of materials flow. However, as we discussed in the last section, the planner normally cannot use intensity of flow alone as the sole measure. We must not lose sight of handling difficulty and *cost*. Handling cost and speed of delivery are two of the true objective determinants in developing optimum plant layouts. Although in some cases we make the simplifying assumption that intensity of material flow is directly related to handling costs, that is obviously not always the case. *Equivalency* has to be taken into account. To further illustrate, consider one fork truck movement of 6,000 lb (2,722 kg) of sand in one large container from point *A* to point *B*. Although the total weight moved would be the same, this *is not* equivalent to the manual movement of 120 50-lb (23-kg) bags of materials, moving the same distance, one bag at a time, from point *B* to point *C*. The material handling labor costs associated with the manual movement is much higher. The costs of the fork truck and its fuel, power, and maintenance notwithstanding, I leave it to you to conclude whether *A* or *C* should be closer to *B*.

In complex manufacturing environments, it is sometimes difficult and extremely time-consuming to determine actual materials handling labor costs without the aid of specialized computer programs. If we are fortunate enough to have simple cart movements or fork truck movements between activities, then we may only need to count movements to gage the intensity of material flow and approximate costs. Other mixed materials handling equipment situations are not as easy to analyze. Also, in the initial stages of designing a totally new plant or new layout, we need to calculate pure materials flow intensity or volume first, because we have not as yet determined the proper materials handling equipment to be used.

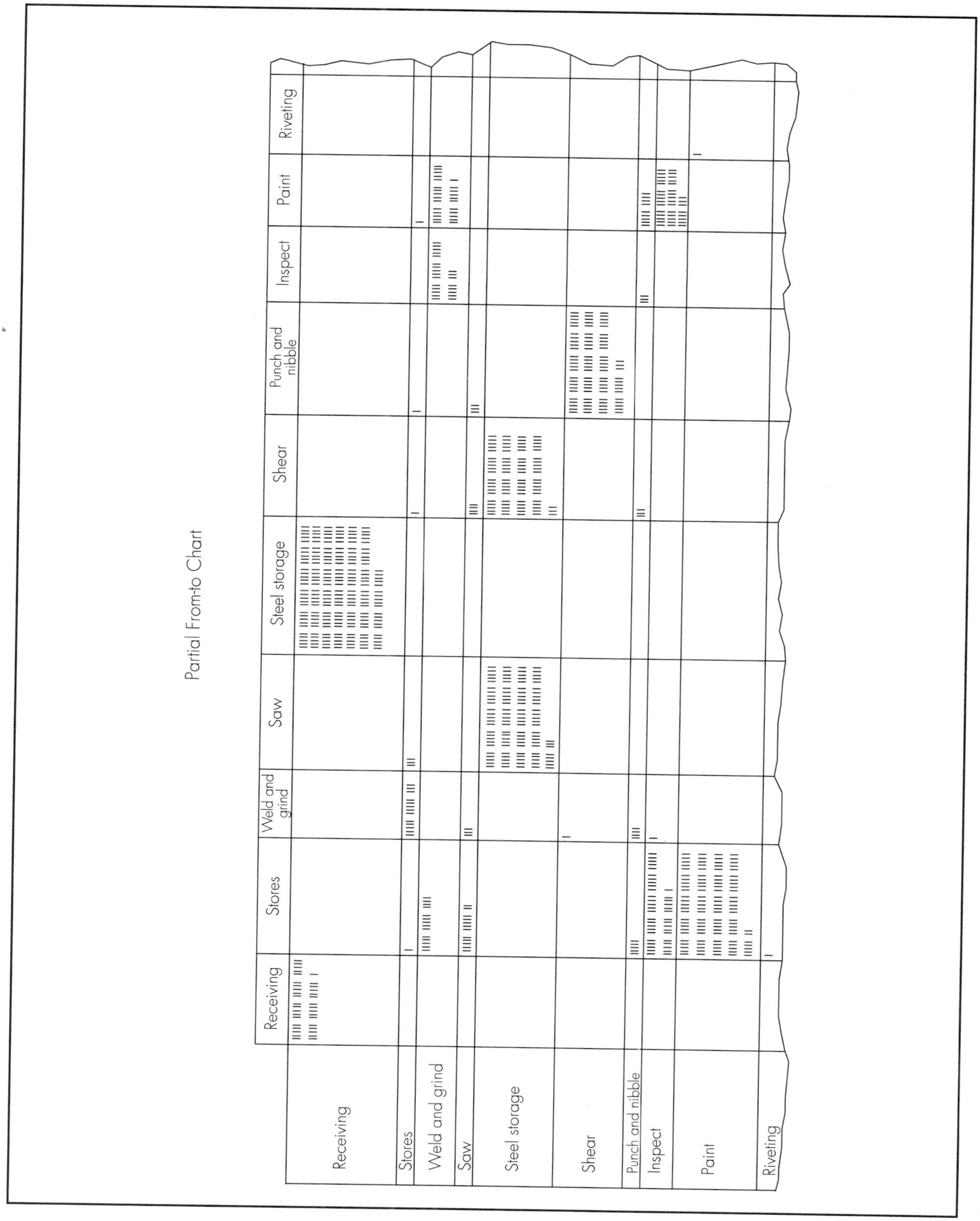

Partial From-to Chart

	Receiving	Stores	Weld and grind	Saw	Steel storage	Shear	Punch and nibble	Inspect	Paint	Riveting
Receiving	‖‖‖ ‖‖‖ ‖‖‖ ‖‖‖ ‖‖‖ ‖‖‖									
Stores		‖ ‖‖‖‖ ‖‖‖								
Weld and grind		‖ ‖‖‖‖ ‖‖	‖‖‖‖ ‖‖‖‖ ‖‖							
Saw		‖‖‖‖ ‖‖‖ ‖	‖							
Steel storage			‖	‖‖‖‖ ‖‖‖‖ ‖‖‖‖ ‖‖‖‖ ‖‖‖‖ ‖‖‖‖ ‖‖‖‖ ‖‖						
Shear			‖	‖‖‖‖ ‖‖‖‖ ‖‖‖‖ ‖‖‖‖ ‖‖‖ ‖	‖					
Punch and nibble		‖		‖‖‖‖ ‖‖‖‖ ‖‖‖‖ ‖‖‖ ‖‖‖‖ ‖	‖	‖‖‖‖ ‖‖‖‖ ‖‖‖‖ ‖‖‖ ‖‖‖‖ ‖‖				
Inspect		‖‖‖ ‖‖‖‖ ‖‖‖‖ ‖‖‖ ‖‖‖‖ ‖‖‖‖ ‖‖‖ ‖‖‖‖ ‖	‖			‖		‖‖‖ ‖‖‖‖ ‖‖‖‖ ‖‖‖ ‖‖		
Paint		‖‖‖ ‖‖‖‖ ‖‖‖‖ ‖‖‖ ‖‖‖‖ ‖‖‖‖ ‖‖‖ ‖‖‖‖ ‖‖‖ ‖‖‖ ‖‖				‖	‖‖‖ ‖‖‖‖ ‖‖‖‖ ‖‖‖	‖	‖‖‖ ‖‖‖‖ ‖‖‖‖ ‖‖‖‖ ‖‖‖‖ ‖ ‖‖‖‖ ‖‖‖‖ ‖‖	
Riveting		‖							‖	

Figure 5-19. A simplified from-to chart provides rough volume and routing data.

Often the operations listed on the from-to chart are taken directly from operations lists or process routing sheets for each item or a sampling of items manufactured. The paths between activities with materials flow are taken directly from operations process charts. While a total plant layout might follow the product from materials received through pre-processing materials storage to final shipment, few lists—and consequently, few from-to charts—show this entire sequence of events. They seldom indicate movement of scrap, rework, or auxiliary materials and supplies movement. They typically make no mention of empty container and dunnage removal, in-process storage stops, diversion to weigh scales, or inspection and packaging. Yet each of these operations requires movement that must be considered in establishing the relationships among operations. For this reason, some floor observations are absolutely required for situations where items are currently being produced.

As mentioned, we should try to list our operations in their natural move sequence down the left-hand side of our chart. If we have a straight-through process flow, most move data will fall above the dark diagonal line in Figure 5-18. If we are working with an existing plant which does not have an optimum layout, more than a few flows will probably show up below the diagonal line. A poorly laid out plant will show many flows below the diagonal. Flows below this line usually indicate backflows of materials moving against the natural forward flow.

If there is not enough time to do a thorough detailed analysis of all materials moves within a plant, this analysis process is sometimes shortened by sampling, i.e., from-to charts are established for the parts comprising the top 15 or 20% of the total annual quantity (or those products with the highest profit margins produced)—the fast movers. The reason for this type of sampling is because this top 15 or 20% usually accounts for most of the materials handling problems and costs. This will be demonstrated when we rank these materials flows.

The importance of a good materials flow analysis in establishing flow-based relationships cannot be stressed enough. Too often backflows and dunnage flows are entirely overlooked. Conversely, when materials flows are heavy,

awkward, or difficult to move (other than in a straight line, such as in a steel mill), the process usually dominates the layout. A from-to chart is normally not required in those cases.

When less than a dozen or so products or parts need to be charted, it is usually best to make an individual process chart for each. When more need to be charted, it is sometimes best to use a multiproduct chart such as shown in Figure 5-9.

The multiproduct chart is useful in visually analyzing the ideal flow pattern within the plant. If several backflows are noted, it immediately points out the potential need to reverse the order of the two locations or work cells. One can change the operational location sequences on the left-hand side of the chart until he or she has a minimum of backflows. The objective is to have the maximum intensity of materials flow move in a straight line. Obviously, such an analysis or rearrangement would not necessarily be needed in a chemical plant or cosmetics plant where materials may be piped to their destinations.

When a large number of items must be tracked, it is wise to first set up a consolidated operations process chart such as shown in Figure 5-20. This chart shows the generic movement of materials and is a great help in reducing the complexity of major layout projects. Even on simpler projects, some professionals use this type of chart to quickly become familiar with the overall scheme of plant operations.

As just one example, I was responsible for completing a materials handling analysis within 80 days on a 250-acre shipbuilding yard with more than 5,000 employees and literally thousands of materials movements. More than 800 activities on site were reduced to fewer than 50 activity areas for materials flow and even fewer for the consolidated process chart (Figure 5-21).

FACTORS AFFECTING ALL MOVES

Four basic factors affect all moves. They can easily be remembered as the "4 Ms."

1. **Materials.**
2. **Methods** of handling.
3. **Moves** — transport distances and elevations.
4. **Money** — cost of the manufacturing philosophy, materials movement, and handling equipment.

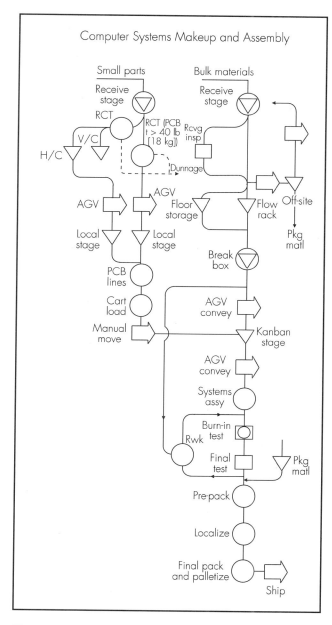

Figure 5-20. For tracking a large quantity of items, a consolidated operations process chart such as this one helps cut the complexity of major layout projects.

Given a particular manufacturing philosophy, we can normally change only factors 2 and 3. We might also be able to help ourselves by changing the package or container within which our materials (factor 1) reside or are transported. The fourth item, money, is determined by how

well we do with other factors. Factor 3 is certainly a large component of costs, but it usually is not the most important factor affecting *total* enterprise costs.

You probably have noticed by this point that plant layout and materials handling are intertwined and inseparable. The bulk of the discussion here is focused on factor 3 — optimizing and shortening moves and transport distances.

Manufacturing and materials control strategies, focused product layouts versus focused process layouts, inspection needs and certified suppliers, point-of-use storage and *kanban* stocking practices, JIT, and agile and lean manufacturing philosophies are all comprehensive *cost-related* subjects of their own. All of these factors and philosophies affect total enterprise costs.

One needs to be wary of falling into the common industrial engineering trap of believing that a new plant layout will solve all of a company's problems. In some poorly-managed companies, implementing a new layout would be akin to rearranging the deck chairs on the Titanic. In most companies, materials handling costs, while generally a large component of total costs, cannot compare to the inefficiencies associated with poor manufacturing philosophies.

As just one example, I recently worked on a materials handling logistics project in a large plant approximately 1,640 ft (500 m) long. There was a management debate over whether to have two separate receiving areas, at opposite ends of the building, to serve each (approximate) half of the plant. The two separate areas would involve additional personnel for the receiving docks and incoming inspection, but the total transport and materials handling labor would be far less (shorter transport distances). Although our simulation showed total materials handling costs would be lowered significantly with the two receiving areas, additional personnel costs could potentially offset a large portion of the savings. A much larger cost savings, as well as a cycle-time reduction, could be achieved by implementing a certified supplier program and eliminating the need for a large incoming inspection department. To make these transport-cost types of calculations and decisions, a quantitative computer analysis is required. We address this in Chapter 13.

Yard Activity Areas—Materials Movements
800 Activities Consolidated to 46

Area no.	Area name
1.	Warehouse no. 1 main purchased parts storage and government furnished material
2.	Warehouse no. 2 stock material storage and government furnished material
3.	Warehouse no. 3 stock material, hardware, pipe fittings, door plant material, and surplus material
4.	Pipe storage (RAW) stock pipe, job pipe, surplus
5.	Marshalling/staging area—pipe details
6.	Yard fab pipe storage
7.	Drum storage/electric cable storage
8.	Fab part storage/pre-fab unit steel storage
9.	Sheet metal shop and associated storage
10.	Main pipe shop
11.	Main machine shop
12.	Pipe module shop
13.	Carpenter shop and associated storage
14.	Paint shop (near carpenter shop)
15.	Fab shop #2/blacksmith
16.	Wet dock #1
17.	Wet dock #1 shops
18.	Wet dock #2
19.	Wet dock #3
20.	Lower building and launch ways
21.	Main upper dry dock
22.	Plate shop
23.	Beam line building
24.	T-beam building
25.	Pickling plant
26.	Rubber plant
27.	Blast/paint/area K
28.	Structural blast/paint
29.	Platen #1
30.	Platens #2/#3
31.	Upper building ways—area 307, platens #4 and #7
32.	Platens #5/#6
33.	Platens #8, #9, and #10
34.	Platens #11, #12, and #13
35.	Platens #14, #15, and #16
36.	Platens #17, #18, and #19
37.	Platen #20
38.	Platen #21
39.	Platen #22
40.	Platens #23 and #24
41.	Platen #25
42.	Platen #26
43.	Paint storage
44.	Roto-blast
45.	Blast house

> See Figure 5-10 for the Consolidated Operations Process Chart

NOTE: The main steel storage area, rod storage, paint storage, and maintenance are not included.

Figure 5-21. This real-world example shows clearly how consolidated operations process charts can reduce the complexity of layouts.

6

CALCULATING SPACE REQUIREMENTS

GENERAL CONSIDERATIONS

Unfortunately, there is no magic formula for calculating space requirements. However, there is an axiom that most experienced facilities planners agree on: *manufacturing managers in large companies tend to overestimate their individual cell or departmental space needs.* This appears to stem from their always getting somewhat less than what they originally requested in money, time, capital budgets, expense budgets, etc. As a result, if a manufacturing manager in a large company absolutely knows he or she needs, say, 5,000 ft^2 (465 m^2) of new space in an expansion, the manager will typically ask for 7,500 ft^2 (700 m^2) or more. So let the plant layout planner beware. This tendency to overstate space requirements needs to be taken into account and double checked using the planner's personal evaluation of existing conditions. Space requirements cannot and should not be determined by a single person. Also, space requirements should be determined quantitatively wherever possible. Avoid using subjective interviews as a total basis for determining space.

On the opposite end of the spectrum, there is a propensity for inexperienced planners and managers to underestimate space requirements for support functions and nonproductive areas. As discussed earlier, this is particularly true in the case of main aisles in a manufacturing plant. Depending on the overall size of the facility, it is not uncommon for the main aisles of a plant engaged in light manufacturing to account for anywhere between 10 to 18% of total underroof floor space. This does not include the aisles in-

side departments or cells, only those aisles connecting the major activities. Figure 6-1 and Tables 6-1 and 6-2 show the proportional space used for main aisles in a typical plant.

As demonstrated in Tables 6-1 and 6-2, one should be wary of blindly quoting percentages for main aisle space. The percentage depends on the base number used in the calculation. Planners should take note of the difference between making calculations based on net space and gross space. These two measurements can be confusing unless the base on which they were calculated is understood first.

When all of the area requirements for the activities are established, the planner has to multiply the total net space by a main aisle factor. This calculation yields an approximation of the total underroof space required.

USE OF GROSS BUSINESS RATIOS

A useful measurement or yardstick for helping to determine gross space requirements may be developed through the use of space and revenue ratios. Also, employee projections can be very useful. Keep in mind that using space and revenue ratios is not an exact method and is used for approximations only.

A typical space ratio is based on the current number of employees a particular plant or department may have. Plant-wide ratios or ratio ranges are generally available or can be calculated. (Sometimes this data can be estimated for very large companies, such as those in the electronics and automotive industries, using publicly available information.) The range of

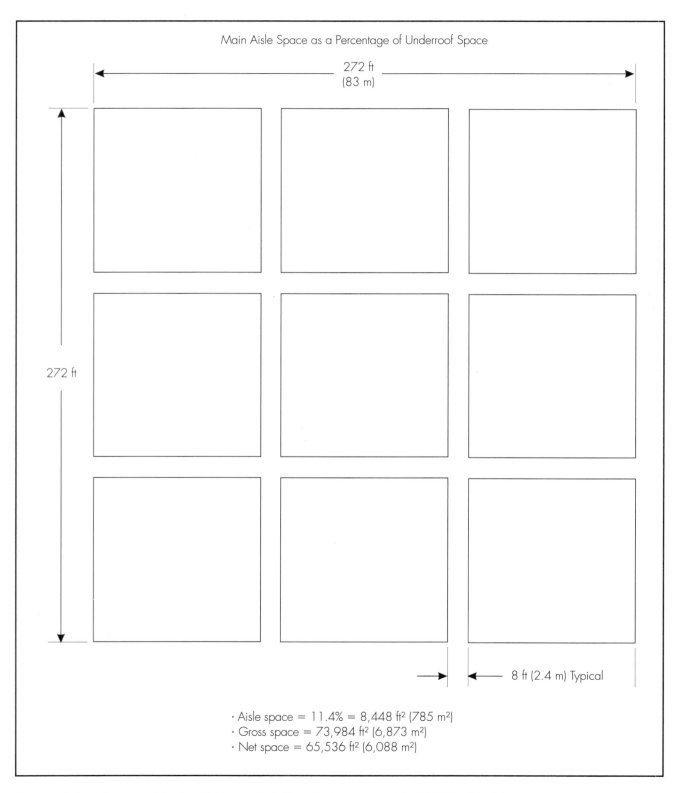

Figure 6-1. *Just the main aisles in a light manufacturing plant can account for 10-18% of total floor space.*

Table 6-1. Calculating Percentage of Underroof Space
Main Aisle Space Worksheet

Building east-west length (ft/m)	400/122		
Building north-south width (ft/m)	250/76		
Number of east-west aisles	2		
Number of north-south aisles	2		
Aisle width (ft/m)	8.50/2.60	10.00/3.00	12.00/3.66
Gross area (ft²/m²)	100,000/9,290	100,000	100,000
Aisle area (ft²/m²)	10,761/1,000	12,600/1,171	15,024/1,396
Total work area (ft²/m²)	89,239/8,290	87,400/8,120	84,976/7,894
Number of work area blocks	9	9	9
Average work area/block (ft²/m²)	9,915/921	9,711/902	9,442/877
Average east-west block length (ft/m)	127.67/38.91	126.67/38.60	125.33/38.20
Average north-south block width (ft/m)	77.67/23.67	76.67/23.36	75.33/22.96
Aisle/gross area	11%	13%	15%
Aisle/net work area	12%	14%	18%

Table 6-2. Calculating Percentage of Underroof Space
Main Aisle Space Worksheet

Building east-west length (ft/m)	100/30		
Building north-south width (ft/m)	100		
Number of east-west aisles	2		
Number of north-south aisles	2		
Aisle width (ft/m)	8.50/2.60	10.00/3.04	12.00/3.66
Gross area (ft²/m²)	10,000/929	10,000	10,000
Aisle area (ft²/m²)	3,111/289	3,600/334	4,224/392
Total work area (ft²/m²)	6,889/640	6,400/595	5,776/537
Number of work area blocks	9	9	9
Average work area/block (ft²/m²)	765	711	642
Average east-west block length (ft/m)	27.67/8.43	26.67/8.13	25.33/7.72
Average north-south block width (ft/m)	27.67	26.67	25.33
Aisle/gross area	31%	36%	42%
Aisle/net work area	45%	56%	73%

variance between different companies is relatively small in office environments but tends to expand in manufacturing areas. These ratios are obviously also dependent on the type of manufacturing being performed. A typical measurement would be the gross number of square feet (meters) of underroof manufacturing space per manufacturing employee.

On the revenue side, a typical ratio is sales dollars per square foot (meter) of plant space per employee. These types of gross ratios tend to be similar for similar plants within an industry.

One must be careful to adjust all business activity-dependent ratios for plant capacity/activity. For instance, in a recession, a plant may be operating with 80% of its normal complement of employees. If we were to use this temporary low number of employees to calculate space ratios, the space per employee ratio would be overstated and an adjustment would be required. Obviously, the number of work shifts also needs to be taken into consideration.

Space-related ratios can be very useful. For example, if you are planning a plant expansion and your current factory is congested, you could first calculate the current number of square feet (meters) of underroof space per production worker (total production space divided by the total number of production employees). You could then establish that ratio as the *worst* condition you would ever want to see again. In other words, that would be the ratio you would expect to have just prior to the *second* expansion cycle (the expansion *after* this current one you are planning) say, in 5 or 6 years from now. Of course this ratio would only be a valid measurement if the nature of the business does not change and no major technological changes are foreseen. If technological changes are expected, the result could be factored into the ratio.

In addition to a total overall ratio, you also need to break down subratios for the various different classifications of space as we will see in the next section. For the moment, however, let's assume we have a relatively simple plant with only five or six space ratios.

We also need to prepare a listing of our company's sales history. We can then coordinate with the engineering, sales, and marketing departments to develop a sales revenue projection for the next 5 years or so. An example projection is shown in Figure 6-2. Based on current and historical data, we should also construct a projected trend line and ratio for the sales revenues generated per employee. Figure 6-3 shows a sample trend line. Using this trend line and the sales revenue projections, we can estimate the number of manufacturing employees in, say, 5 years when we would expect to expand the plant again.

At this point we have two key pieces of information. One, we have the current manufacturing space ratio per employee that we know is "tight." Two, we have an estimate for the number of manufacturing employees that will most likely be required in 5 years to support the projected sales levels. If, in fact, we expect to reach the same level of "tightness" in 5 years, when we expect to expand again, we can simply multiply the 978.5 ft² (90.9 m²) per manufacturing employee times the projected increase in number of manufacturing employees 5 years out. That will give us a gross estimate (or a double check) of the extra space needed for today's construction. Of course, this figure would have to be adjusted for the growth in other areas. We also need to make an estimate of *the proportional main aisle space required.*

Obviously, we are discussing very gross estimates. However, a company should continuously monitor its space *productivity.* Just as labor hours per product shipped is a valid measurement of productivity, so is square feet or meters of plant space employed per unit produced.

Before finalizing gross projections, some detailed space analysis should be performed and compared to the gross estimates to raise the probability of success. It is always better to reach an estimate working from two different sets of data, in this case, both gross and detail space workups.

SPACE BALANCE FOR LONG-TERM PROJECTIONS

As mentioned in the previous section, space ratios can be a valuable tool for estimating space requirements for the current and next expansion. These ratios also can be modified and used for making projections for longer-term needs.

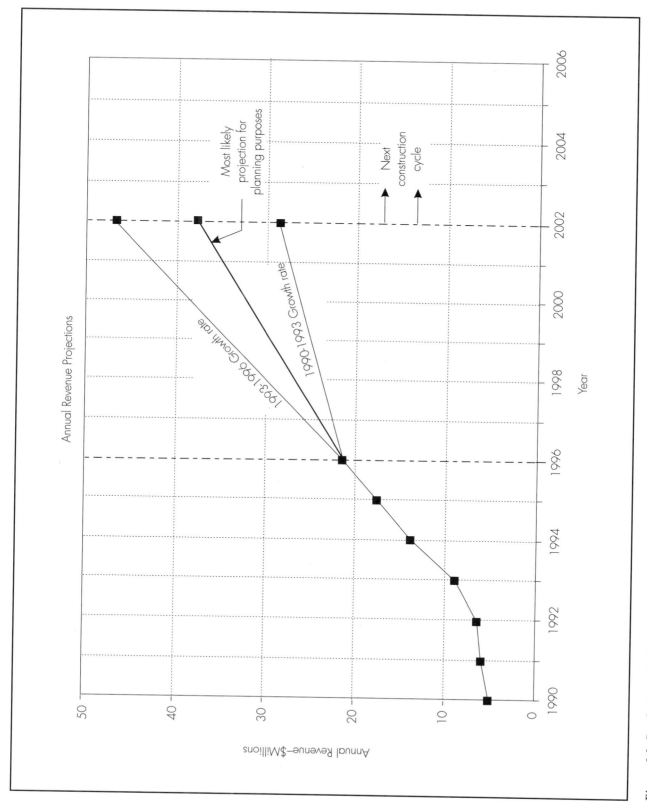

Figure 6-2. By plotting sales history and company business trends, space needs can be projected.

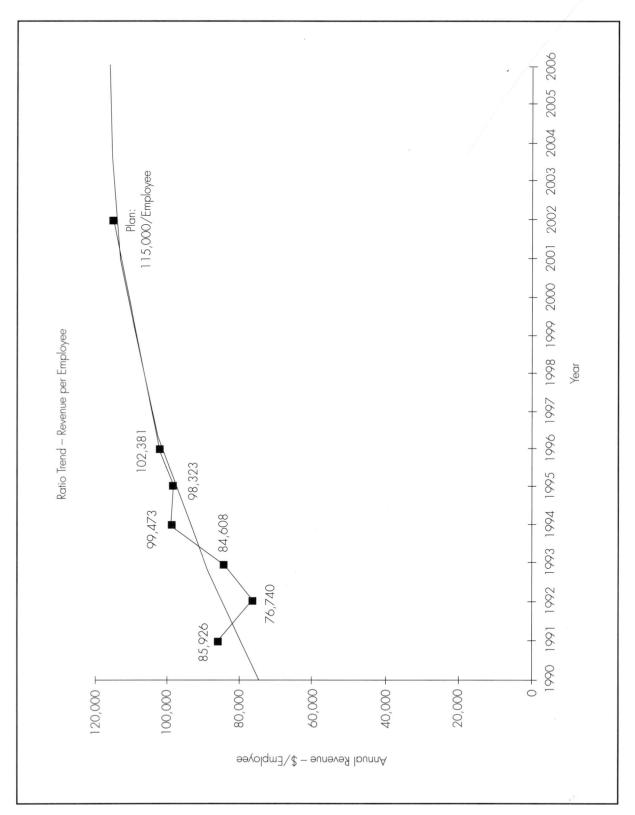

Figure 6-3. *Using the overall revenue projections of Figure 6-2, a revenue per employee projection can easily be calculated for a given time frame in the future. From these trends, an estimate can be made of total employees needed for the projected time frame.*

For estimating long-term underroof space requirements, it is usually wise to perform a space balance analysis. This analysis classifies all underroof space needs into 5 to 10 separate classifications. The space classifications most often used are:

1. Primary operations,
2. Secondary operations,
3. Inspection and test areas,
4. Storage areas (may want to categorize),
   a. Raw materials,
   b. WIP materials,
   c. Finished goods,
5. Service and support areas,
6. Shipping/receiving dock areas,
7. Offices,
8. Main aisles.

Existing space is normally catalogued on a spreadsheet. A percentage *space balance* is then calculated, based on all of the floor area being used. It is quite helpful if the company has a historical record of these relative proportions of space, especially if the company has been in the same general type of business for many years. Companies that keep such records can track productivity in relation to space usage and develop intelligent master plans for future expansions and new facilities. A typical space usage record, in a somewhat different format, is shown in Figure 6-4.

The important point to remember is that all underroof space must be allocated and accounted for. In other words, *all* of the company's space should be classified into one of the categories previously discussed. When all of the percentages are added, they must total 100% of the space being used.

These same or slightly modified percentages may be used to project the facility block layout at what is called the site saturation level.

SITE SATURATION/MASTER PLANNING METHOD

Site saturation is a term used to signify what an industrial site will look like when it is completely saturated or filled to its maximum usage configuration. When a company has reached its site saturation point, no additional building additions or expansions are possible (unless they rise vertically). If we have developed our space projections accurately, the site saturation plan may be used to predict the maximum production output that can be achieved at a particular site. Admittedly, this is only a crude and gross estimate, but it is a far better estimate than a pure guess.

The site saturation plan actually defines the framework of a long-term *master facility plan*. The construction configuration outline plan will show the relationships between the major categories of space via a simple spatial relationship diagram analysis (see Figure 6-5). The plan will also take into account all access and egress roads, easements, setbacks, building configuration(s), unusable space, parking, etc. Frequently, the terms "master facility plan" and "site saturation plan" are used interchangeably.

The site layout at the saturation point can be developed in a number of complex ways, but many of these are beyond the scope of this text. The simplest way to develop the plan is to first develop a long-term space balance, as we discussed in the last section. The next step is to develop a relatively simple relationship diagram showing the major blocks of space in relation to one another. The blocks of space are then drawn into a building shell in the same proportions as we developed in our predicted space balance. The future space balance would be adjusted for those technological and other changes that are predicted to occur. Keep in mind that the proportions of all categories of underroof space still need to add up to 100%. At this point, we have an estimate for the percentages of space that will make up the future plant in our master site saturation plan, but we have not yet determined the actual areas. An iterative analysis is required to determine how large a *total plant* we can have on the site.

The analysis is done best with a spreadsheet model similar to the one shown in Table 6-3. Parking areas alone can be equivalent to 40% (or more) of underroof manufacturing space in some industries. Parking area(s) plus building areas need to be iteratively calculated and summed until all available site space is used. Several trials are probably required. The calculation is somewhat difficult since parking areas are a function of the number of employees (and work shifts); the number of employees are a

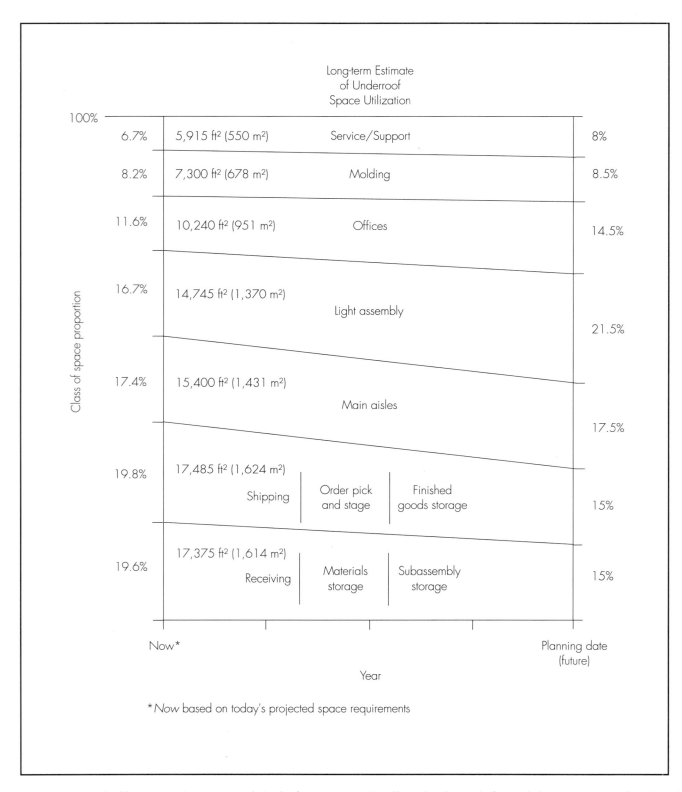

Figure 6-4. Invaluable to projecting space needs in the future is a continually updated record of space balance by primary function.

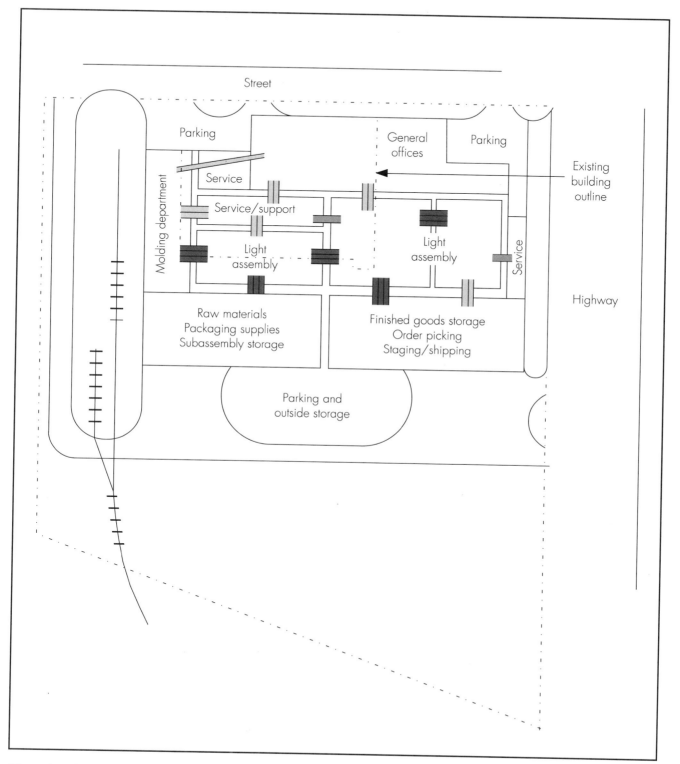

Figure 6-5. A master facility plan incorporates a construction configuration outline plan showing maximum facility size at site saturation.

Table 6-3. Site Saturation Space Projections

	Now (ft²/m²)	Plus 4 Yr (ft²/m²)	Plus 8 Yr (ft²/m²)
Receiving	5,872/546	4,228/393	4,932/458
Shipping		2,819/262	3,288/305
Fabrication	6,454/600	7,809/725	8,971/833
Welding	6,644/617	7,906/734	9,434/876
Paint	7,398/687	8,878/825	10,209/948
Assembly	5,630/523	6,869/638	7,544/701
Machine shop	4,630/430	5,649/525	6,482/602
Raw material storage	6,180/574	7,478/695	8,714/810
WIP stores	5,554/516	6,665/619	7,720/717
Finished goods storage	6,654/618	7,985/742	9,316/865
Offices	6,530/607	7,771/722	9,273/861
Main aisles	8,252/767	9,943/924	11,492/1,068
Parking	26,523/2,464	32,759/3,043	34,081/3,166
Total underroof space	96,321/8,949	116,759/10,847	131,456/12,210

function of production output, which in turn is a function of both space, equipment, and employee productivity. Keep in mind we can make predictions of all of these factors as outlined in the prior section.

A rational site layout or master plan can be projected, based on an existing space balance, building code requirements, productivity estimates, dominant consideration analysis, etc. Of course, this analysis assumes that the company's type of business does not change to any great degree over the length of time of the space balance projection. Those portions that we are certain will change have been allowed for in our space balance. Even though we can estimate the space configuration for a master plan, in most cases it is more difficult to predict when the company will actually reach the site saturation point.

One who is wary of master plans should keep in mind that a site saturation plan is not cast in concrete and inviolate. The site saturation plan is a dynamic document, that should be updated on an annual or at least biennial basis. Done correctly, it is one of the most valuable planning tools a company can use.

A site saturation plan should be in place *before* we make any short-to-intermediate facility expansion plans. I have experienced several situations where the development of the master plan helped prevent a whole series of badly planned departmental moves and removes. These moves and removes would have taken place several times in a short time span. Without a master plan, short-term thinking will normally result in long-term consequences costing the organization greatly.

EQUIPMENT UTILIZATION CONSIDERATIONS

A key factor in estimating space is equipment utilization. It is extremely important that we use the correct utilization factors in space planning. On new plant designs in particular, the use of incorrect utilization factors or ignoring utilization completely is a common error made by many inexperienced planners and architects. The effects of errors in utilization cannot be overstressed since these types of errors can be almost impossible to overcome. Errors in utilization and, therefore, in space projections, will form the root of problems for a very long time, usually until the company's next expansion. In some situations, space needs can be severely underestimated. Let's consider a hypothetical case.

Suppose an equipment vendor guarantees that its equipment system (six pieces of different linked machines per "system") will produce at a rate of 88 pieceparts an hour and the catalog brochure indicates the equipment occupies 1,075 ft² (100 m²). Further assume that we have a daily production need of 2,100 pieceparts per day, 7 days a week, and that we plan to operate the equipment on each of three daily 8-hour work shifts. Neglecting aisle space for the moment, how much area do we need for the equipment?

One could take the simplistic view that we need 2,100/3 (or 700) pieceparts per shift, divided by 8 hours per shift. That equates to 87.5 parts per hour. Since the equipment system guarantees 88 per hour, we only need one "system" or 1,075 ft² (100 m²) of space, right? Absolutely *wrong!*

What are some of the fallacies with this analysis? First, it would be impossible to plan for a 100% utilization level for the equipment. There must be some downtime for both planned and unplanned maintenance. Second, even though the vendor guarantees a production rate of 88 pieceparts per hour, that guarantee implies that all pieces of equipment are properly maintained and operating. If we assume that each different piece of linked machinery is randomly down 5% of the time for repairs, a simulation would show that the expected rate of production would be only 64.7 pieceparts per hour. The analysis would show that we would actually need 1.36 systems without any further safety factor for labor inefficiencies such as late starts and early quits, personal fatigue and delay time, etc. Instead of one system, in fact we would need two systems, and our space and equipment requirement would be *double* what had originally been planned.

We could improve the situation by adding a small buffer inventory between each of the six linked machines. Ideally, the buffers would be sized to allow most of the machines to continue operating until the downed machine is repaired. Of course, to accumulate buffers, some flexible overcapacity must be designed into the system, as it should be. The ideal buffer quantity size would be just approaching zero or full depletion as the errant machine comes back on line.

The building of buffer inventories goes against the grain of most of today's management "experts." Unfortunately, many of these gurus have spent far too much time in the classroom and far too little time on the shop floor in a *real-world* production operation. You will find that the most efficient production systems in use today make good use of point-of-use buffer inventories, albeit very sparingly. The key to success is the right sizing of the buffers, which trades off buffer inventory carrying costs (which can be a significant portion of annual part/product value) and potential idle labor costs.

Utilization errors also can arise in the planning of existing plant expansions. Consider the following example.

- Our current machine has run historically at 65% utilization when fully scheduled and is producing 500 pieces per day. That is, the 100% rating (like the 88 per hour rating in our previous example) is 769 per day, but we have never been able to achieve that.
- Our projected expansion requirement is 1,300 pieces per day.
- Management is assuming better control with the expansion and wants to plan on an 85% utilization level.

Question: How many machines are required to produce 1,300 pieces per day?

The total number of machines required is:

$(500 \times 0.85) \div 0.65 = 653$ pieces/day

$1,300 \div 653 = 2$ machines needed (with a negligible amount of overtime).

However, if we have planned our space requirements on an 85% utilization level and fall short of reaching that target, *we will not have the capacity or space required to reach the 1,300-piece-per-day output requirement.* If we are only able to achieve the historical 500-piece-per-day machine output, *we will need three machines, not two,* to reach our production requirement. This underestimate is more likely to occur in plant expansion examples such as this one. Why? Everyone expects productivity improvements in plant expansions, and rightly so. No one expects to remain with the status quo. Upper management may push for such an improvement and an 85% utilization level sounds reasonable. However, in most situations, unless there is some *significant physical change* in the way a company does business or in its mode of manufacturing, it is almost impossible to go from a 65% utilization rate to an 85% level with minor changes to a layout. Such an improvement is certainly possible but it usually requires many more system changes.

The planner needs to be wary of wishful thinking creeping into the planning process when developing equipment utilization factors. Certainly, everyone wishes they could move from a lower utilization rate to a higher one. However, it takes much more than wishful thinking to achieve improvements on the order of 85/65—that is equivalent to more than a 30% improvement in productivity. One should always question such assumptions, unless there is a factual basis for them. The best way to do this is to determine space and layout requirements *using actual production history as a calculation basis* and then visually compare the results to the requirements using the projected new utilization rates as a basis. This "what-if" space differential comparison should be reviewed by all involved with approving the final layouts. With layout drawing comparisons, everyone involved can see the impact of utilization changes on space and equipment needs.

ADJUSTING TODAY'S NEEDS

The method most frequently used for calculating gross expansion space needs for existing products and processes is commonly called *conversion* or the *converting* method. This approach also may be used for designing a new plant that will be using processes and equipment similar to an existing plant. The conversion method usually entails a spreadsheet type of data tabulation, as shown in Figure 6-6. (We will use it later in the development of block or macro plant layouts.)

Before beginning any space extension calculations with the conversion method, all of the existing underroof plant space usage must be tallied. All blocked activities are listed, similar to the space balance discussed earlier, but in a much more detailed fashion. Most manufacturing plants have anywhere from a dozen to 40 or so major activities or "blocks" of space. Since initially we are developing macro layouts, entire departments or manufacturing cells are normally listed as activities, e.g., receiving, raw materials storage, assembly, painting, etc. The current space used by each activity is listed in the first column of Figure 6-6. It is very important that all space, *including main aisles*, is accounted for. The conversion method builds upon this initial space listing, so it should be as accurate as possible. A total area should be calculated based on the existing building shell. The space used for main aisles also should be calculated (not estimated) from existing layouts. The difference between the building shell area and the space taken by main aisles should be completely accounted for in the sum of all of the areas listed for activities. Generally, it takes several checks and double checks to ensure that the numbers are correct (or within a small percentage of measurement error).

The next step of the data-gathering process used in the conversion method can be either very simple or highly complex. If no major process changes are contemplated, it can be a relatively simple process.

In the simple "rough estimate" case, a small group of individuals experienced with the plant processes performs a walk-through and observation of each activity area and operation. The team observes what is happening, speaks with individual operators, takes notes, and generally records pertinent data regarding the "looseness" or "tightness" of space. Preferably, the core team consists of the group leader or shop-floor supervisor, accompanied by at least one experienced industrial or manufacturing engineer. In

		Space Planning Worksheet (Conversion Method)			
No.	Activity Name	Existing Area (ft²/m²)	Percentage Adjustment	Adjusted Area (ft²/m²)	Percentage of Total
1	Drum storage	272/25.3	30%	354/32.9	0.5%
2	Compressor	150/13.9	0%	150/13.9	0.2%
3	Shear room	507/47.1	0%	507/47.1	0.8%
4	Saw room	1,105/102.7	100%	2,210/205.3	3.4%
5	Production office	683/63.5	60%	1,093/101.5	1.7%
6	Machine shop	3,700/344	90%	7,030/653.1	10.9%
7	Strain gage and wiring	1,190/110.6	90%	2,261/210.0	3.5%
8	Pregage	121/11.2	90%	230/21.4	0.4%
9	Heat treat/inspection	320/30.0	130%	736/68.4	1.1%
10	Cutup/setup	438/40.7	0%	438/40.7	0.7%
11	Receiving	500/46.5	125%	1,125/104.5	1.7%
12	Stockroom	1,289/119.8	60%	2,062/191.6	3/2%
15	Pressure assembly/test	2,600/242.0	175%	7,150/664.2	11.0%
16	Electronics assembly	990/92.0	150%	2,475/229.9	3.8%
17	TIG weld	700/65.0	150%	1,750/162.6	2.7%
18	Lunch	1,047/97.3	90%	1,989/184.8	3.1%
19	Shipping	576/54.0	160%	1,498/139.2	2.3%
20	Administrative office	1,733/161.0	25%	2,166/201.2	3.4%
22	Restrooms	600/55.7	100%	1,200/111.5	1.9%
26	Engineering offices	2,651/246.2	20%	3,181/295.5	4.9%
27	Sales/marketing	3,470/322.4	110%	7,287/677.0	11.3%
30	Engineering lab	1,562/145.1	20%	1,874/174.1	2.9%
31	Model shop	483/44.9	40%	676/62.8	1.0%
32	Drafting	331/30.7	300%	1,324/123.0	2.0%
34	Computer room	120/11.1	0%	120/11.1	0.2%
102	ACC assembly	380/35.3	0%	380/35.3	0.6%
200	DCT	850/79.0	100%	1,700/158.0	2.6%
13a	Load cell assembly/test	1,103/102.5	175%	3,033/282.0	4.7%
13b	RMA	350/33.0	100%	700/65.0	1.1%
23b	Boiler room	200/18.6	−50%	100/9.3	0.2%
	Subtotal	30,021/2,789.0		56,799/5,276.7	
	Main aisles	3,603/335.0	120%	7,926/736.3	12.2%
	Total	33,624/3,124.0		64,725/6,013.0	100.0%

Figure 6-6. In planning for expansion or new-plant space requirements, most planners use the conversion method, starting with current usage (adjusted for today's conditions) by functional area and adjusting upward to projected needs.

some cases, representative operators from the individual activities are included as *ad hoc* members of the team. It has been my experience that the smaller the "roving" core team, the better. It should not exceed four people, lest there be too much confusion, resulting in poor communication.

Some plants have many teams, each comprised of individuals within each work cell. I feel this multiteam approach should only be used *after* results are obtained by the initial core team. Space estimates need to be done by someone with an overall, unbiased point of view to coordinate information.

There may be one or two additional "go-arounds" to confirm or correct estimates after the individual work-cell teams give their inputs. In the simple case, gross estimates of excess space or additional space needed to meet *current* production needs are recorded. In other words, if it appears that a particular activity is too congested at current production levels and 5% additional space would relieve the congestion, then the +5% figure would be recorded in the adjustment column. Conversely, if the team is in, say, a storage area that is using too much space for current operations, a negative percentage might be included in the adjustment column. The results of the walk-throughs and consensus space adjustments are recorded and tallied.

As you have likely noted, unless there is an experienced industrial or manufacturing engineer on the team, simple "fixes" or potential minor changes to improve conditions can be overlooked. For example, the company may be using a counterbalanced forklift truck in an operation where a stand-up reach truck would suffice. As we shall see in a later section, the use of a stand-up truck could potentially reduce the aisle width (for load turn-in space) by a significant percentage. Similarly, an experienced industrial engineer could notice some wasted time and motion in the observed processes and recommend some simple labor or materials savings improvements. It pays to have an experienced individual on the team. Usually, but not always, the operators themselves do not have the breadth of experience required to recognize simple technological changes that could impact their space needs. (Many operators do, however, have a better vision than an outsider on fundamental changes that would improve their operation.)

The next step is to record any known upcoming changes in space. For instance, a new piece of equipment may be planned for a particular activity and space may have to be added for that. Some other technological change may be planned that will have an effect on space. A plan to use more or fewer work shifts may have an effect on space, and so on. These "new-activity" space adjustments will need to be made and recorded in the next column of our conversion worksheet.

All of these adjustments, plus any change in the proportion of major aisle space, need to be tallied. In the typical completed conversion sheet shown in Figure 6-6, the difference between the total of all adjusted space and the current existing space is the minimum new space needed for the expansion.

The conversion method basically gives the planner a chance to adjust the current plant area to what it should be today. It is far better to make future capacity adjustments and extensions against a revised, accurate base than, say, against a very congested base area. For example, suppose that next year we want to double the output of a particular 1,075 ft² (100 m²) department which is operating 24 hours a day, 7 days a week and has no extra capacity. Perhaps we may even duplicate the department in another building. A walk-through of the existing department indicates it is too congested and could use between 5 and 10% (let's use 8%) more space, just to meet current production needs. Assume we do not have enough time or resources to perform a detailed analysis of space needs. How much total space should we allow for the department at a doubling of capacity?

Using the conversion method:

1,075 ft² (100 m²) $\times$ 1.08 $\times$ 2.00 = 2,322 ft² (216 m²) would be required.

Be forewarned that, in most situations, the space required for increasing capacity or production levels is not a simple linear function of production quantities. A much more detailed and thorough analysis may be required.

Although, in the simple case, the conversion method uses a somewhat subjective or approximate approach to estimating space, this does not necessarily limit the procedure. If the planner wishes to do a more extensive and thorough space needs analysis, he or she can still make use of the basic adjustment process. The actual calculation of space needed for equipment will be much more detailed.

DETAIL DETERMINATION/ CALCULATION OF SPACE NEEDS

The method most frequently used for determining work center *specific* space needs is called the *rough layout* or the *production center* method. Typically, plan view templates of each piece of production and storage equipment used within the work center are prepared (to scale). These can be prepared manually or via computer-based images. Each access door of each piece of equipment is shown fully open, with any special access noted (e.g., long shaft pulls for maintenance, etc.). Approximate spacing between each piece of equipment is established (see Figure 6-7). Usually, it is not necessary to know the actual detail placement of each piece of equipment, although we should be establishing relative sizes and aspect ratios of the spaces at this time (e.g., a particular work center may need to be several

times longer than its width, etc.). Detail placements come in a later project phase, after we have developed a macro layout.

The purpose of this space determination phase is to make sure we account for everything within the work centers, as well as all of the supporting indirect areas, so that we do not short-change ourselves on the space allocations. It is especially important that we also take into account material pickup and dropoff points, materials staging areas within the work cells, etc. Even though the final detail layouts may change considerably, we should be able to estimate the block of space required at this time. Frequently, a combination of the production center layout and the conversion method is used to develop space needs.

PLANNING STORAGE AREAS WITHIN THE PLANT

Planning the gross or detail space for storage areas within a manufacturing plant usually requires a thorough analysis and some early decisions on equipment choices. Equipment selection can have a tremendous impact on storage area space requirements.

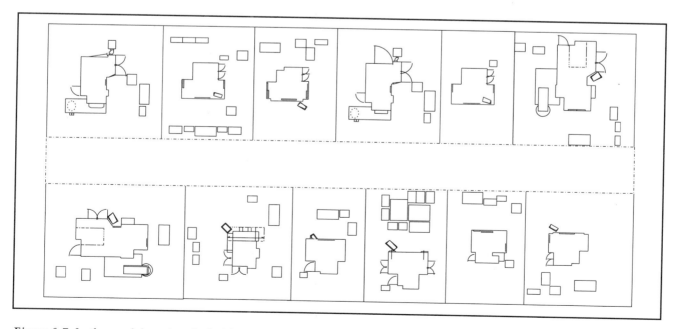

Figure 6-7. In the rough layout method of departmental layout, approximate spacing between each piece of (drawn-to-scale) equipment is established, ensuring that everything that needs to be in the work center fits in the space allocated.

Suppose we are planning to use selective pallet racks for storage of pallet loads of materials within the plant. The aisle spacing within the storage area is dependent on the forklift truck type that we plan to use. Numerous industrial trucks are available for internal factory and warehouse use. For pallet storage, typical aisle widths for various trucks are shown in Table 6-4.

Figures 6-8 and 6-9 show plan views of several storage modules and the floor space required per pallet. The middle graphic on each figure assumes that pallets are stacked four high. Different stacking heights will, of course, yield different results.

As you will note, using a stand-up reach truck, instead of a typical counterbalanced truck, will save a considerable amount of floor space. Keep in mind, however, that there are pros and cons associated with all industrial trucks. For example, a counterbalanced truck is much better suited for truck dock work than its stand-up, small-wheeled cousin. Other aspects of materials handling equipment are discussed in further detail in a later chapter.

Generally, it is more space efficient to run pallet racks in the long direction rather than the short direction within a rectangular storage area. Sometimes companies install racks in the short direction for reasons other than space efficiency, such as to provide quick location access with short aisles for a manufacturing operation or shipping area abutting the "long" dimension. Both conditions can be dynamically simulated in specific cases to determine the best alternative.

Another important factor to consider in laying out pallet racks is the avoidance of space-inefficient, standalone, single-bay sections of a rack. Figure 6-10 shows a 50% reduction of WIP inventories in a plant where the number of pallet racks was reduced. Due to all-around area constraints, that particular relayout was quite inefficient. The 50% reduction in pallet inventories achieved less than a 40% reduction in floor space.

Some companies elect to suffer the materials handling penalties of a double-deep rack installation to save floor space (see Figures 6-11 and 6-12). Although double-deep racks make good use of space, usually it is the last-choice selection for manufacturers requiring total selectivity for stored loads. To get access to a back pallet, the front pallet must be removed and placed on the floor (or elsewhere) nearby. Then, the pallet one really needs is removed and also placed on the floor nearby. Next, the first unwanted pallet is picked up and placed in the open back position (which generally requires an inventory location update). Finally, the operator can go back and get the correct pallet and move it to where it is needed. Even with this extra materials handling

Table 6-4. Warehouse Materials Handling Equipment
Typical Aisle Widths

Type	Order picking (ft/m)	Turning (ft/m)
• Counter-balanced ride lift truck	12/3.65	15/4.57
• Counter-balanced stand-up lift truck	9–10/2.74–3.04	12/3.65
• Straddle lift truck	7–8/2.13–2.43	10/3.04
• Reach lift truck	8–9/2.43–2.74	12/3.65
• Double-reach lift truck	8–10/2.43–3.04	12/3.65
• Turret truck	Load width plus 15–28 in./38–71 cm Typical 60–66 in./152–168 cm	15/4.57
• Swing mast	5–7/1.52–2.13	15/4.57
• Stacker crane	Load depth plus Typical 60 in./152 cm	18–20/5.49–61.0 (Maintenance area)

NOTE: Aisle size is for 42 × 48-in. (187 × 122-cm) pallet

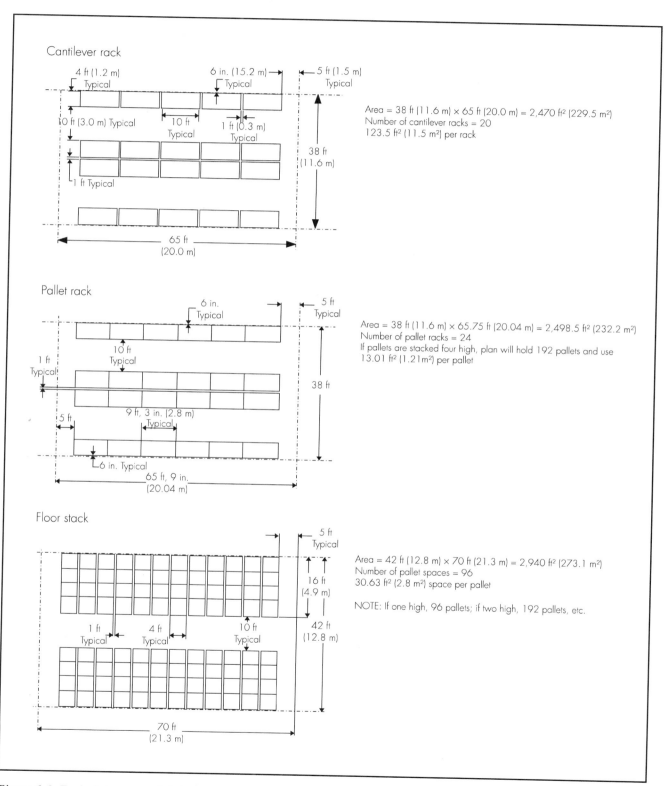

Cantilever rack

Area = 38 ft (11.6 m) × 65 ft (20.0 m) = 2,470 ft² (229.5 m²)
Number of cantilever racks = 20
123.5 ft² (11.5 m²) per rack

Pallet rack

Area = 38 ft (11.6 m) × 65.75 ft (20.04 m) = 2,498.5 ft² (232.2 m²)
Number of pallet racks = 24
If pallets are stacked four high, plan will hold 192 pallets and use
13.01 ft² (1.21 m²) per pallet

Floor stack

Area = 42 ft (12.8 m) × 70 ft (21.3 m) = 2,940 ft² (273.1 m²)
Number of pallet spaces = 96
30.63 ft² (2.8 m²) space per pallet

NOTE: If one high, 96 pallets; if two high, 192 pallets, etc.

Figure 6-8. Typical storage module layouts using a sit-down counter-balanced lift truck.

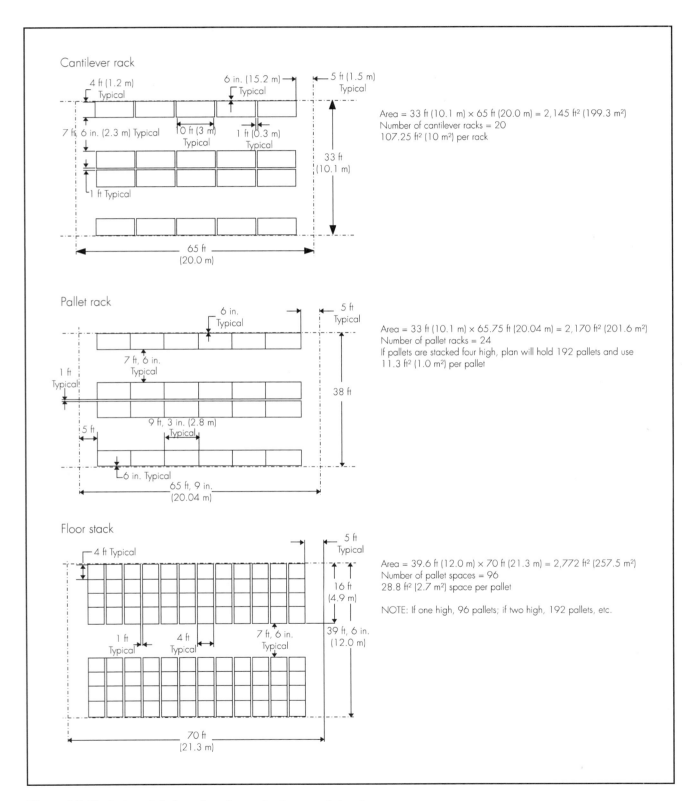

Cantilever rack

Area = 33 ft (10.1 m) × 65 ft (20.0 m) = 2,145 ft² (199.3 m²)
Number of cantilever racks = 20
107.25 ft² (10 m²) per rack

Pallet rack

Area = 33 ft (10.1 m) × 65.75 ft (20.04 m) = 2,170 ft² (201.6 m²)
Number of pallet racks = 24
If pallets are stacked four high, plan will hold 192 pallets and use
11.3 ft² (1.0 m²) per pallet

Floor stack

Area = 39.6 ft (12.0 m) × 70 ft (21.3 m) = 2,772 ft² (257.5 m²)
Number of pallet spaces = 96
28.8 ft² (2.7 m²) space per pallet

NOTE: If one high, 96 pallets; if two high, 192 pallets, etc.

Figure 6-9. Typical module layouts using a stand-up reach truck.

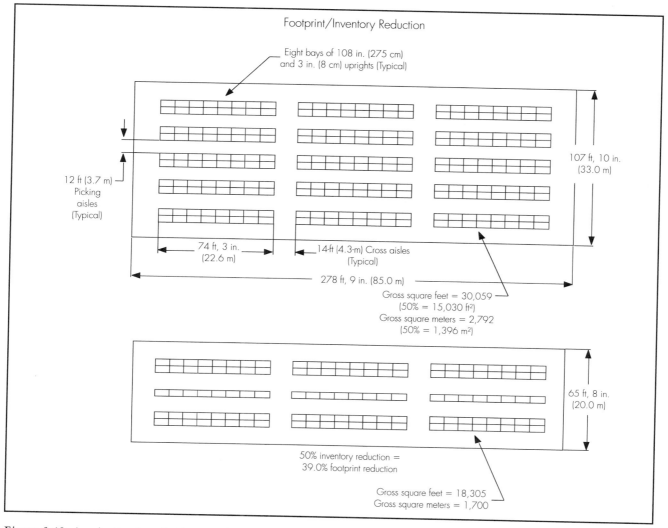

Figure 6-10. A reduction in pallet inventories does not necessarily generate a one-for-one reduction in gross storage area.

and added chance of error (first pallet placed in a different position than recorded), many companies will suffer the frustration and added labor costs to save space. Conversely, some manufacturers that always store and access two pallet loads of a particular batch (such as in pharmaceutical manufacturing) can find double-deep racks to be a very efficient storage choice.

Whether using single-deep or double-deep racks, you can gain additional space by bridging the cross aisles as shown in Figure 6-13. Make sure, however, that fork truck drivers are trained to lower the truck masts before entering a cross aisle, lest he or she creates another aisle!

Some of the other automated storage aids that can help save space and improve inventory accuracy and accessibility include unit load automatic storage and retrieval systems (ASRS), miniload storage and retrieval systems, horizontal and vertical carousels, and vertical power columns. Typical examples of these types of systems are shown in Figures 6-14 through 6-18.

Some companies will use carousels and power columns for WIP banks connecting two or more work centers. Frequently, this is done with horizontal carousels assisted by pick and place cranes, Figure 6-19. As shown in Figure 6-18, power columns can offer even more flexibility.

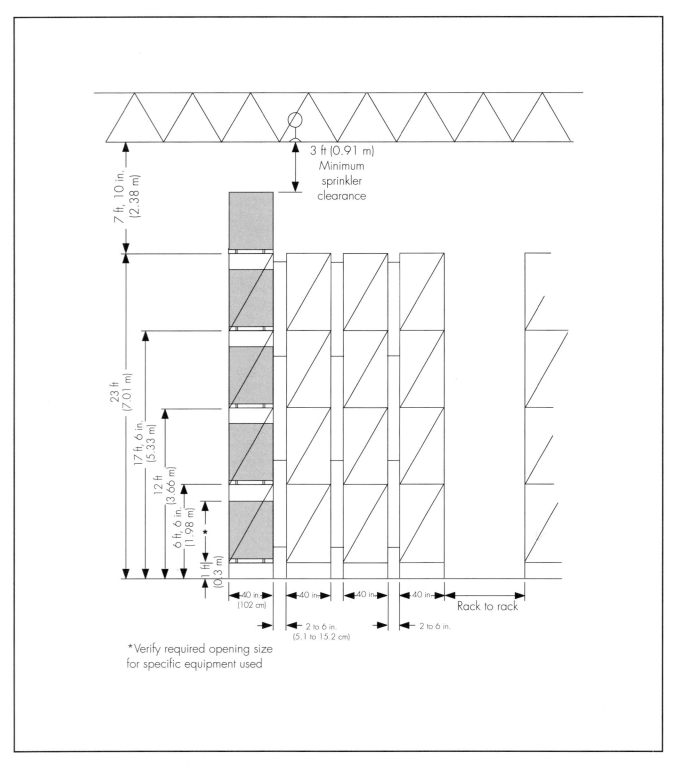

Figure 6-11. *Although double-deep racking saves floor space, there are penalties in the form of repeated materials handling, unless load selectivity is not a factor.*

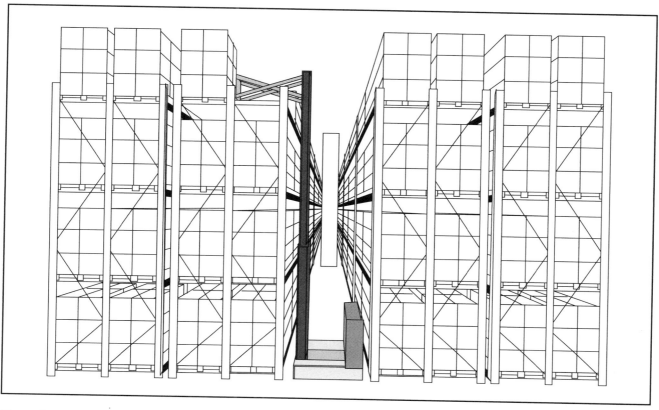

Figure 6-12. *Even with a double-reach sideloader, double stacking requires a pick-move-pick-remove-pick-replace sequence of handling, as opposed to a pick-move sequence in single racking.*

They can serve simultaneously as both a WIP storage bank and as a vertical transporter to or from a second processing level.

A particular manufacturer's philosophy will frequently play the most important role in storage equipment selection. I have been involved with some plants where the policy is to install completely automated, high-cost, state-of-the-art storage equipment whether cost-justified or not. In those plants, image may be more important than low cost and low inventories. Conversely, I have seen plants where, for no technical reason but only because of company policy, no fork trucks were to be used within the manufacturing areas. No pallet rack could be used and only hand pallet jacks were allowed. By policy, (and in one case, an *intentionally* placed physical height barrier) no floor stacks of material could be over two pallets high. Obviously, this type of tight inventory policy can force JIT methods,

good materials control procedures, and low shop-floor inventories. Some would say this policy is what lean manufacturing is all about. Others would say it is the cause of excessive employee stress. It might also result in the leasing of outside storage space (or complete outsourcing of logistics, resulting in loss of jobs or benefits for existing employees). In fact, each situation is different and needs to be viewed on its own merits.

Similarly, management of some firms have policies that discourage the use of turret trucks (used in very narrow aisle storage situations), since they cost approximately four to six times more than standard counterbalanced forklift trucks and have somewhat higher maintenance costs. Even though the additional height capability and floor-space savings (in pallet rack situations) of the turret truck may cost-justify their implementation, some management teams are

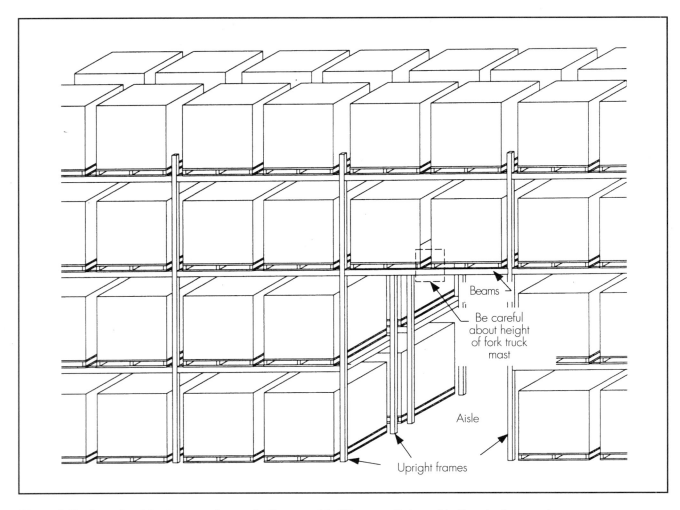

Beams

Be careful
about height
of fork truck
mast

Aisle

Upright frames

Figure 6-13. Over-the-aisle storage makes productive use of facility space that would otherwise be vacant.

still reluctant to approve their purchase. One typical reason is that the counter-balance truck gives more flexibility if inventories were ever reduced to the point where high racking would be empty.

One of the most critical factors needing a decision before we can finalize factory and activity space requirements is whether or not a point-of-use storage philosophy will be used within the plant. Many companies that have what I describe as *critical care* inventories, such as those in some pharmaceutical plants, aerospace plants, and high-tech electronics plants, cannot have total point-of-use storage systems. Traceability, accountability, value, quarantine materials, government regulations, and several other factors may prevent total or even partial point-of-use storage. For most other companies, *controlled* point-of-use storage systems are generally recommended.

Many companies already have consumable materials delivered and stocked internally in their plants by vendor personnel on a chit or blanket-order purchasing arrangement. Some tooling companies even provide and maintain their supplies in vending machine types of dispensers directly on the factory floor. In those situations, the vendor has the responsibility for managing the inventory and seeing that supplies never run out.

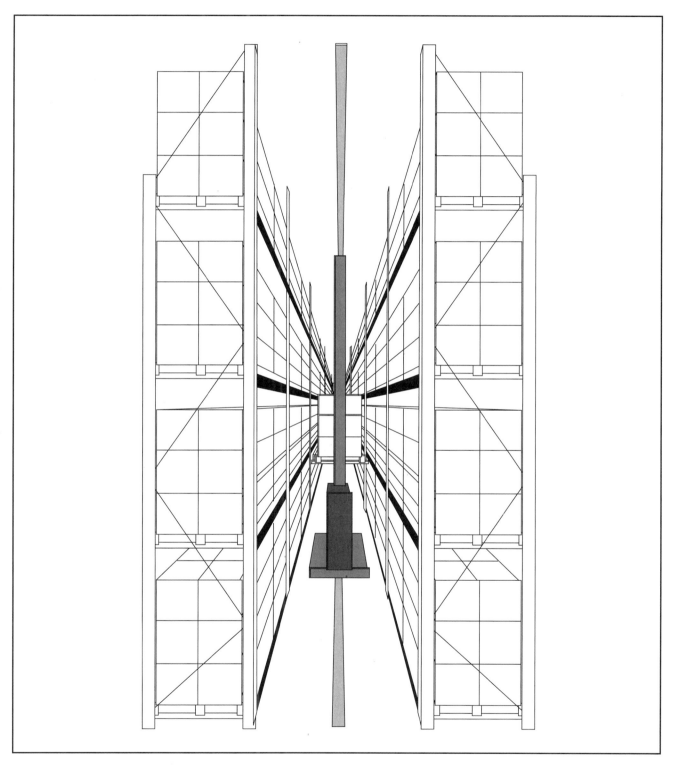

Figure 6-14. *Automated storage and retrieval systems can be designed for virtually all types of storage. This one, for large pallet loads, requires much less operating space than an industrial truck.*

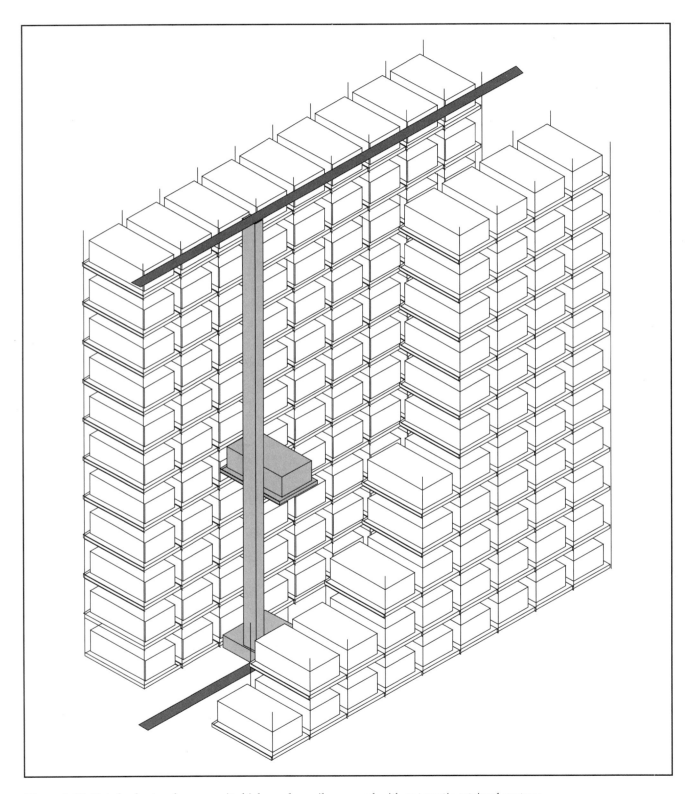

Figure 6-15. Tote loads stored many units high can be easily accessed with automatic retrieval systems.

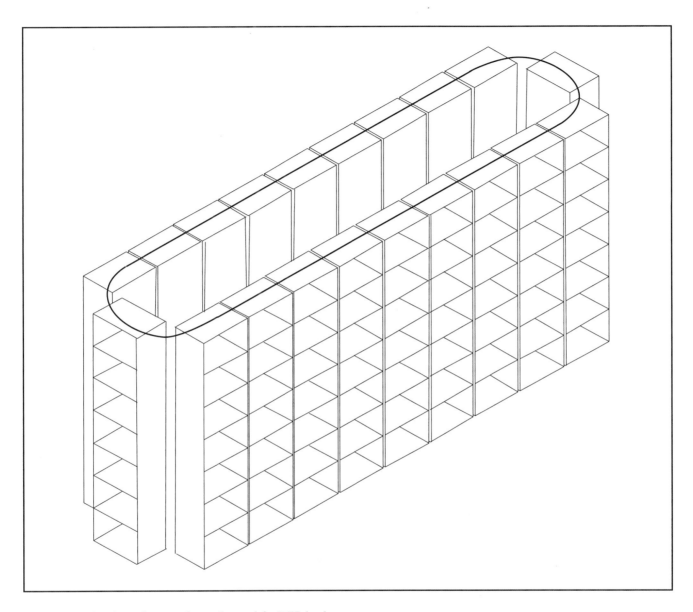

Figure 6-16. Horizontal carousels can be used for WIP banks.

Often, changes in materials storage philosophies are extremely difficult to implement. Habits developed through years of traditional centralized storage situations have a tendency to build up a cadre of people whose jobs are dependent on maintaining the status quo. Because materials storage and staging space is such an important component of total factory space, these change decisions need to be made as early as possible in the layout planning process.

For small parts storage within a plant, shelving systems are commonly used. A typical in-plant shelving unit is shown in Figure 6-20. One must be careful to take ergonomics and shelf-cube utilization into account in sizing and specifying shelving systems. Many companies do not permit storage on some shelves, particularly the lower ones. Back injuries from excessive bending to reach parts on lower shelves can be a serious problem.

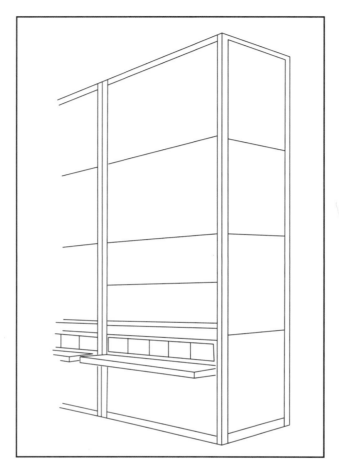

Figure 6-17. Vertical carousels bring materials to the worker at the appropriate work height.

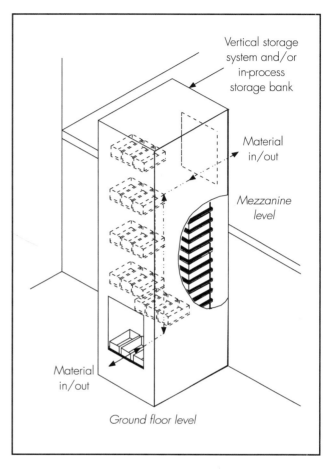

Figure 6-18. Vertical power columns are ideally suited to operations in which materials are received or stored at one level and accessed for production or delivery at another.

Of course, if lower shelves are not used, shelf-cube utilization is much lower and more shelving units are required. Even normal shelves must be sized at a relatively low utilization level. Empirical studies performed by my firm show typical cube utilization in most shelving installations to be between 25 to 40%. That is to say, on average, a 12-in. (30.5-cm)-deep shelf 35 in. (89 cm) long with a shelf-to-shelf spacing of, say, 8 in. (20.3 cm) has 3,360 in.^{3} (55,060 cm^3) of space available (12 in. × 8 in. × 35 in.). Empirical measurements of actual factory cube storage on shelves such as this range between 840 and 1,340 in.^{3} (13,765 and 21,959 cm^3) of parts or products stored per shelf. Some cases have been much lower.

In some factories, the increasing number of short runs or model changeovers has created horrific logistics problems in getting materials to the right place at the right time. The Harley-Davidson motorcycle assembly plant in York, Pa., in the U.S. is a prime example. Motorcycles are typically built to a customer or dealer's order. Almost every other motorcycle coming down Harley's paced main assembly line has a different paint scheme. The line is very closely packed with little room between workstations. Manual pushcarts are used to bring painted parts to a large staging area near the main assembly line. The materials handling logistics of getting the correctly painted and pin-striped part families (fenders, fuel tanks, etc.) on carts and

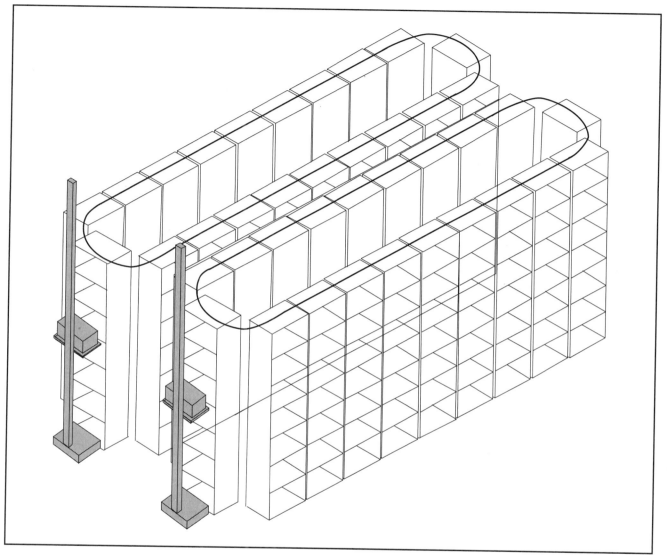

Figure 6-19. *Pick-and-place cranes working in conjunction with horizontal carousels provide an efficient solution to supplying two or more work centers.*

to the line at the right place at the right time for the right motorcycle is an extremely difficult task. The painting logistics alone, while trying to maintain minimum scratch-free inventories, is a huge effort in itself. Logistics aside, Harley's products are in such great demand worldwide that all of their dealers have been placed on an allocation system. At the time of this writing, Harley's assembly plant is running at full capacity and the delivery lead time for a new, custom-painted Harley "hog" is more than a year.

Similarly, I have seen furniture factories where complete furniture sets are fabricated in highly efficient manufacturing cells. Each customer-ordered set may contain widely varying styles of furniture, e.g., sofa, ottoman, chair, love seat, etc. The typical older engineering approach would be to specialize labor and production lines so that chairs are produced on one line, sofas on another line, etc. However, that type of layout does not foster the same level of product quality or labor efficiency as a multifaceted cell

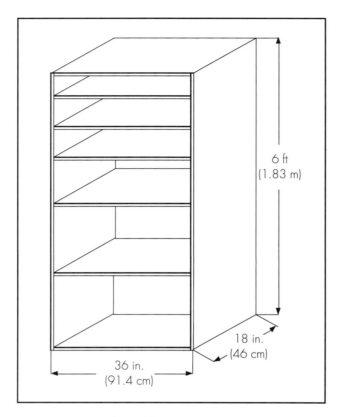

6 ft
(1.83 m)

18 in.
(46 cm)

36 in.
(91.4 cm)

Figure 6-20. Though shelf-storage is perhaps one of the most popular methods of small parts and materials storage, ergonomics and actual shelf-cube utilization are critical factors affecting its efficiency.

approach. The specialized line approach results in the need for a complex scheduling and materials control system and a large storage or staging area to accumulate items as they come off the various lines at different times. The cell approach offers tremendous advantages over the specialized line approach.

Customer-ordered *complete sets* are fabricated within each cell. This ensures fabric matches on all pieces. It also ensures that complete customer-ordered sets stay together, without an extensive storage and consolidation area. The scheduling problem is dramatically reduced. Workers are cross-trained to perform all operations and the labor balance is achieved automatically through the use of an individual and team-based incentive system. Complete furniture sets come out of the cell and, following a quality check (the team gets demerits and must repair any quality

defects uncovered), the sets move directly into a shipping truck.

As with the motorcycle example, the major consideration and problem in furniture "set" manufacturing is the logistics of getting the correct materials, precut wood parts, and fabrics to the line at the right place at the right time for the right furniture set. It is being done, but it is not an easy physical task. Similar to the cart staging in the Harley-Davidson example, innovative materials handling carts and movable/rotatable shelving systems are required for maximum efficiencies.

In electronics manufacturing and in other manufacturing lines where very small parts are used, both sophisticated and low-tech storage and delivery systems are used successfully. These range from motorized horizontal and vertical carousels to seemingly ageless Rotobins™ on movable carts. Rotobins are basically multitiered, manually-rotatable, "lazy-susan" types of shelving devices. They have been used for more than a half century in maintenance departments for storing various nuts, bolts, and other hardware. They are now being used on wheeled carriers in high-tech assembly areas for quick model changeovers on assembly lines.

Kanban *Considerations*

In the industrialized western countries, the Toyota JIT approach has been widely adopted, particularly in the automobile manufacturing industries (which includes suppliers to the major assembly plants). In many companies, the traditional MRP scheduling systems have been modified (but not abandoned) to include JIT considerations. That is to say, lot quantities have become smaller, and the number of deliveries have increased proportionally to the lot size decreases. However, many companies are still building to a "push" or projected schedule, albeit with smaller lot sizes, smaller overall inventories, and an increased number of transactions.

In a true "pull" type of production system, frequently called "demand flow" or "demand pull," parts are only pulled, manufactured, and delivered as needed, based on actual customer demand. (Obviously, this system assumes that someone in the chain of supply carries a stock of items that have very long lead times.) The

Toyota or Japanese approach to aid the pull manufacturing system is commonly called *kanban* manufacturing. *Kanban* incorporates a very simple visual scheduling and replenishment approach.

Most newcomers to manufacturing believe the visual replenishment system originated in Japan. However, similar visual replenishment systems had been in use in the U.S. and U.K. for more than half a century prior to the mid-1960s. These, in fact, were the most widely used systems prior to the advent of computerized materials control systems (MRP and MRP-II). The unique difference that the Japanese emphasized was the absolute requirement for minimum "front-end," WIP, and finished goods inventories. The Japanese emphasized building as closely to actual demand as possible and not to a long-term forecast. Although visual production replenishment systems had been in place for many years in the west (in lieu of more formalized scheduling systems), the Japanese drive to force inventories to an absolute minimum was missing. Also, the "pull" versus "push" manufacturing rationale was not in the forefront of western thinking prior to JIT and the *kanban* system. For many years, the industrialized west was essentially performing materials handling and replenishment based on visual cues; however, the main stockrooms were still being replenished on a scheduled push basis to a longer-term production forecast. High inventory levels obscured many efficiency problems within manufacturing operations. Smaller inventories require a much more focused manufacturing management approach.

Instead of a formal production scheduling system, the *kanban* system uses a visual communication signal that automatically notifies material handlers that parts are required. Paperwork and scheduling workloads are reduced significantly. If done successfully, the need for large inventories is also significantly reduced. Of great importance to facility and layout planners, floorspace requirements for inventories are reduced.

In a typical *kanban* system, all materials are received (by computer transaction) and placed either in raw materials inventory or, preferably, delivered directly to the consuming line where the materials reside in a temporary WIP inventory account. If materials are transferred from raw material stores, they are transferred directly to the WIP inventory account. If planned correctly on a *kanban* basis, there is a very small level of WIP inventory, usually measured in several hours' or several days' of supply. There is no further tracking of materials past the replenishment point of the consuming operation. WIP inventory is automatically relieved or backflushed when the product or finished parts leave the production operation and become finished-goods inventory. One of the keys to a successful *kanban* system is maintaining an absolute minimum WIP inventory level.

Another major key to a successful *kanban* system is relying on the shop-floor employees to manage the material replenishment pull system. *Kanban* in reality starts at the finish. When products are completed and parts are consumed to a predetermined quantity, a card, flag, light, or other visual signal (at the point-of-use replenishment location) is used to signal that new replenishment parts are required. Typically the number of parts for the replenishment is predetermined and containers hold certain multiples of received package sizes. More sophisticated systems may use an electronic terminal input to signal the need. If a card system is used, the card in the empty container (or posted on the supply board) shows what is needed, where it comes from, and the point of delivery or usage.

The production operation will typically be supplied with two small containers for each particular material. When one container is depleted, the empty container or *kanban* card serves as the signal to a roving material handler to bring a new container of materials.

There are also dual-card *kanbans* for multiple consumers of a single part or for some processes, such as heat-treating, that require specific large-size lots to be produced.

The sizing of the *kanban* containers is dependent on the length of time needed by the handler to service a number of "calls" and get back with a new supply. Most companies work towards keeping the *kanban* size as small as possible without causing production line shortages. Typical *kanban* sizes are multiples of visual package sizes (e.g., rolls, cartons, pallets, etc.) so established to avoid unnecessary detail counting.

To calculate the *kanban* size for a particular component, we must first calculate the *usage* size per replenishment cycle:

Usage Size = $\dfrac{(Q \times q)\, r}{hC}$

Where: Q = quantity of product produced per work shift

q = quantity of this component used in each product produced

r = replenishment time in hours

h = the number of hours per work shift

C = the number of components per unit container (e.g., number per carton or number per pallet)

For example, suppose we are assembling 3,200 appliances per 7.25-hour shift and one heater assembly is installed at one operator position in each appliance. Also suppose we want an approximate replenishment time of 4 hours and that 900 heater assemblies are supplied in each palletized container. How many palletized containers do we need at the point of use?

The *usage size* calculation would be (3,200 × 1 × 4) ÷ (7.25 × 900) = 1.96. We would round this up to two palletized containers. Therefore, each *replenishment* would consist of two palletized containers. The operator would need an additional container to be working from with the two "empties" stacked for a *kanban* takeaway. Therefore, the *total* required *kanban* line space would be sized to hold *three* containers. If this turned out to be too much space in the layout, we would need to reduce the *kanban* line storage space to only enough room to hold two containers. This of course would require a *kanban* size reduction to one palletized carton with replenishment on a more frequent basis (approximately every 2 hours in this case). The replenishment signal in the revised situation would be one empty container with a *kanban* card or flag.

Needless to emphasize, there is a cost tradeoff among the number of deliveries required, storage space allocations, replenishment labor, optimum package sizes, detrash needs, returnable containers, etc. *Kanban* in general, and JIT techniques in particular, can have a dramatic effect on factory space requirements. These decisions or strategies need to be worked out as early in the planning process as possible.

Importance of Inventory Turns

When planning storage areas, it is very important to have the company set an inventory turn or inventory stocking objective for the materials in each area. It is also important that the company have a clear definition of inventory turns.

Most companies establish inventory turns in monetary units, whereas in space planning, we are more interested in physical units. For the planner unfamiliar with this terminology, we need to convert *physical* inventory turns into how many hours or days of inventory we plan to have on hand in each storage and manufacturing area.

For example, let's suppose a company uses one widget subassembly in each product unit it sells. Let's also assume the company is building 480,000 units per year. The company has 240 production days and wants 240 inventory turns on WIP storage. Basically that equates to 1 day of inventory on hand at the final assembly area or space storing 2,000 widget subassemblies.

However, on checking the production schedule, you find that the company will build 50,000 units in the strongest month of production (20 workdays in the month). If we divide the 50,000 by 20 workdays, we get a 1-day inventory-on-hand requirement of 2500 widget subassemblies. This could conceivably take 25% *more* space than the 2,000 figure previously calculated. The point to be stressed is that the planner needs a clear definition of inventory turns, seasonality effects, and stocking levels *before* establishing the size of storage areas. It is also important that everyone involved with layout approval be aware of the inventory turn objectives. After plant layouts are developed, it can be extremely difficult to change storage space allocations.

Looking at it from another direction, many people incorrectly equate increased transaction levels with a corresponding need for increased space. For example, suppose we have a fixed block of storage space that we cannot change. A good example would be the WIP inentory storage area just discussed. Suppose we have one person servicing the storage area on each work shift. The individuals are stocking and pulling subassemblies and bringing them to the final assembly line. Let's also assume the company has decided to increase production in the coming year by 25%. In an attempt to maintain current space and prevent an expansion, it has been decided to limit the physical inventory on hand to current levels (thereby increasing inventory turns by 25%). In-and-out materials handling transactions will increase by 25% if the mode of inplant handling and delivery does not change. No matter what the inventory turn policy or the size of the area, unless some physical changes are made, more people will be required to handle the increased volume (or the existing people will have to work harder). Since it is difficult to hire 25% of a person on each shift, it is likely that some labor inefficiencies will occur. The main point to be made here, however, is that *increased production levels do not always equate to increases in space, particularly in storage areas*. There is usually a tradeoff in space versus picking and replenishment labor. The planner needs to be careful on this point in interviews with storage area supervisors. Frequently, supervisors and storage area workers equate production increases directly to increased space needs. Obviously, at some point, a physical limit to increasing turns will be reached. As transactions continue to increase without additional storage space, additional people will eventually be bumping into one another and pickup and dropoff staging areas will also need to be enlarged.

Planners should not be left with the impression that increasing inventory turns should not be one of their major goals; it definitely should be. Increasing inventory turns offers very healthy financial benefits to a company. Space savings from increasing inventory turns also can be very significant.

RATIO TREND AND PROJECTIONS

The business ratios we discussed earlier (although still only approximations) can be useful as a "reality" check for determining long-term space needs. It should be stressed that a ratio-trend analysis is based on the company's history. Since history is generally no assurance of future trends, these types of analyses must be carefully reviewed by all interested parties.

The rationale behind ratio-trend analysis lies in computerized regression analysis techniques and the experience curve or learning curve of the company. If enough historical data is available, the planner can track and plot certain ra-

tios related to underroof square footage used. Some of the more common ratios are:

- Square feet (meters) of underroof space per unit produced,
- Square feet (meters) of underroof space per pound or ton of product produced,
- Square feet (meters) of underroof space per kilowatt-hour (kwh) consumed,
- Square feet (meters) of underroof space per production worker.

These values are traced back historically and projected into the future. In the case of kwh consumed or production workers per square foot or meters, these would then have to be correlated with historical and projected production output.

The long-term ratios are then used to project long-term space requirements (see Figure 6-21). Be forewarned that this analysis can become suspect if the nature of the business has changed or there are several physical mergers, etc.

Ratio trend acknowledges that we are all getting better at what we do. In other words, if we are progressing (and staying in business!), we will normally be using less square footage per unit produced today than we were 10 years ago. In many companies, we see a definite trend in plotting this ratio. Of course, owing to periodic technical advances it is normally not as smooth a trend as shown in Figure 6-21. It should be a downward sloping trend, nonetheless.

To get equivalent space comparisons, it is important to adjust historical figures for plant capacity utilization. That is to say, in slow economic times, a company may only be operating at a fraction of its actual capacity. If we were to take a ratio of square feet (or meters) per unit produced during slow periods and use that for our ratio, we would get an erroneous result. All actual production numbers have to be adjusted to reflect capacity and not actual production (which is usually lower than plant capacity). Instead of actual 100% capacity, some companies use an 85% capacity level for standardizing current and historical data.

One of the difficulties with ratio trend analysis is that, although companies keep historical production records, most do not have accurate records on space and corresponding production *capacity*. Usually, only managers involved in strategic planning in major corporations have the

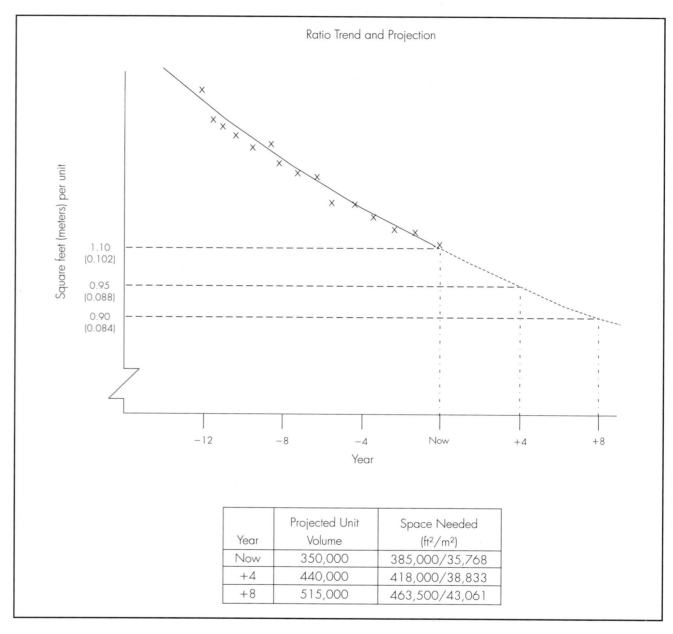

Figure 6-21. *Trends established by plotting the historical track of the company are helpful to the planner, but should not be viewed as absolutes when calculating space needed. Such "ratio trends" and projections should be viewed as aids only.*

discipline and resources to maintain tracking files of this data. Those who do, use the data to develop educated guesses of space needs using long-term marketing projections.

Regardless of whether or not one uses space ratios for trend analysis, it is highly recommended that ratios, such as electrical power us-age per unit produced and per square foot (or meter) of plant space, be maintained. Most companies already track (or should track) ratios, such as labor costs per unit produced. I worked with one large injection molding company that (luckily) kept good records, but did not track this type of information on consumables. During the course of the facilities planning project, we had an opportunity to analyze the historical data. The company was surprised to learn that, on many of its product lines, the cost of electrical power per unit produced was higher than its labor cost per unit. This revelation led to a number of changes, including the purchase of the company's own transformers (to that point it had been leasing them from the power company) and the installation of capacitor banks to improve the power factor (demand) charges applied by the power company. Had the initial electrical power trend analysis not been performed, the company would be at a severe disadvantage in comparison to its competition.

PITFALLS AND REALITIES OF SPACE PROJECTIONS

As I have tried to point out in the previous sections on business ratios and space projections, these methods are approximations at best. Usually, only in new plant designs or in the addition of a special manufacturing process does the planner ever come close to getting the optimum space required. It is much more likely that other factors will have a greater bearing on the allocated planning space.

Strange and ridiculous as it may appear, it is not unusual for many companies to take almost a pure *guess* at the size of the expansion they need. Often, standard building modules of a particular size are proposed by architects or quoted by contractors, and these become the boundaries of the expansion. Typical reasoning for the size selected include generalized statements such as: "We expect to triple the volume of our business over the next 6 years. We will probably need to double or triple the size of our facility." Or at the other extreme: "We will be using certified suppliers and an off-site third party logistics provider who will handle and store all necessary inventories. Therefore, we can triple the volume of our business and we will only need 20% more space." In another case, there may be enough room on the current site to extend the building 100 ft (30 m). In that case, a 100-ft expansion may define the additional space requirement, whether it is too large or too small.

It takes a great deal of ingenuity and innovative thinking to develop optimum layout plans under constraints such as those mentioned. However, this is the challenge to the facility planner. This challenge makes the task more interesting and makes a successful outcome even more rewarding. The third major key to the layout puzzle after materials handling analysis and space needs determination is the subject of the next chapter.

7

RELATIONSHIP/AFFINITY ANALYSIS

THE ENGINEERED PLANNING APPROACH

As discussed previously, with computer assistance, both qualitative and quantitative measurements can be used effectively to compare plant layouts. But not everyone has access to the specialty software programs available for handling these tasks. Currently, by far the great majority of plant layouts created throughout the world are done manually. Here, we discuss some of the typical manual methods used.

The Clone Method

First, there is the *clone* method. Some companies attempt to clone or exactly duplicate an existing plant in a new location. I have even witnessed a facilities manager changing drawing title blocks on an existing building design and cutting out sectional views of prior building construction drawings in an attempt to save architectural fees for a new building. Not only is this practice foolhardy, but it is illegal in most jurisdictions. In addition to violating governmental regulations, in many cases, it even violates copyright laws. Many managers and engineers do not realize that the architect (not the company who hired the architect) usually holds the copyright on architectural design and construction drawings. This is done not only to protect the architect's financial interests but to prevent potential future liability from unauthorized cloning without the architect's and structural engineer's technical review, approval, and consent.

In truth, many companies have cloned individual processes and duplicated them in other plants. Certainly there is nothing wrong with that practice. However, cloning an entire plant (even legally with architectural and engineering help) usually is not a good practice. It may work for some smaller service types of organizations, like restaurants and banks, but for most noncommodity types of processes, it does not result in an efficient operation. Sites and local conditions vary too much. There are always improvements that can be made in layout, access, egress, etc. In my experience, it *always* pays to perform an industrial engineering (IE) and manufacturing engineering (MfgE) analysis *before* committing to a building configuration.

Unfortunately, in more than 50% of the projects I have been involved with, building shells are either a given or have already been determined by others. The IE and MfgE are usually stuck with building constraints that could have been avoided if the "cart was not put before the horse." Some of these constraints will usually add unnecessary manufacturing or materials handling costs for the lifetime of the plant.

The Committee Method

Many companies have also tried using the *committee* method for developing plant layouts. This usually works well for designing individual work cells if the company involves both the work cell operators and an IE or MfgE. Unfortunately, the overall planning of materials logistics to and from the work cells can suffer with the committee approach, particularly if it is a large plant.

This can be prevented if there is an experienced independent and objective person overseeing the total project. Ideally, he or she should have leadership qualities and be intimately familiar with the overall logistics of the receiving and shipping functions and the transport of materials to and from all areas. If this overseer is not in place, the input and output points of the various work centers will invariably not mesh and less than optimum plant layouts and materials handling systems will be developed.

The other major problem frequently encountered with the committee approach is the "paralysis by analysis" effect. This often occurs when all committee members are somewhat passive and there is no clear leadership or timeline-based task schedule. Decision-making can languish for weeks and sometimes months. Committees find it relatively easy to put off making decisions until the "next meeting." I have worked with some companies where engineers and managers spent more than half of their working days in meetings. Unless work was accomplished at the meetings, it did not get done at all. It would not be unusual to have attendees at many of these meetings totally oblivious to what was being discussed. They would be writing or reading extraneous materials picked up in a prior unrelated meeting. Needless to emphasize, poor management practices and a meeting mentality help foster the paralysis by analysis effect.

The Strong-man Method

However, the pendulum could also swing too far in the opposite direction. A company must be careful to not overuse the "strong-man" method. Typically, this is the opposite of the committee approach (although even a committee sometimes has a vocal strong man). It usually occurs when one executive is assigned total responsibility for the project and is not given (or does not want) adequate resources to help plan the project. That person typically has neither the time nor the resources to conduct a quantitative study of the situation and he or she is forced to make subjective decisions. Due to the high rank of this person (or an overbearing personality), these decisions are taken as orders or etched in

stone and are believed by others to be not subject to questioning. This approach simply cannot work effectively on large projects, and sometimes not even on small projects, unless a company is extremely fortunate. Large projects are simply too complex to have one person making subjective decisions. In most cases, the strong-man approach is based on opinion without an objective analysis backing up the view. Invariably, there is little or no documentation of a materials flow analysis supporting the resulting plant layouts. There are times, however, when the strong-man method is unavoidable. The company may think there is no other choice or it is a necessity due to the time constraints associated with new product introductions, etc. Many projects have been completed this way. However, I have never seen one that could not have been improved significantly, had a proper analysis been accomplished.

Straight-Materials-Flow Method

There is also the "straight-materials-flow" method for developing plant layouts. This approach is a well-meaning attempt to streamline product and part flows throughout the plant generally by using a consolidated operations process chart as a guide. Highly experienced individuals familiar with the general sequence of operations of this or other plants may use this approach on small projects. This method is frequently used if the planner does not have enough time to perform a quantitative analysis. Frequently, this is also the starting approach used in the committee method. In small plants with few products, this approach will generally provide acceptable results. Keep in mind that, although the materials may appear to move in a straight-line fashion, the straight-flow approach can sometimes lead to inefficient layouts.

Pitfalls can emerge with the straight-flow approach. For example, in an earlier section, we discussed a plant approximately 1,640 ft (500 m) long. That U.S.-based plant was planning to process (machine) small forgings at the two ends of the plant in two separate business units. With a focused factory strategy, their older product lines were placed at one end of the building and their newer lines were separated and placed at the other end. Since both products required heat-treating, the straight-materials-flow approach employed placed the heat-treating department (a "monument") in the middle of the plant (the short-term total quantity projections for both new and old products were similar). The newer product area had a large sophisticated machining cell installed with underground trenches for chip flow and a complex coolant and chip-treating and processing installation (another "monument"). After an initial layout was completed via the straight-materials-flow approach coupled with "template shuffling" and subjective "opinions," the heat-treating department wound up to be somewhat closer to the old product line than the new product line, but still close to the center of the plant. I was asked to do an overall logistics and materials handling study to determine whether or not the receiving and shipping areas should be centralized or decentralized and to evaluate several other alternatives (fork trucks versus automatic guided vehicles, etc.). Unfortunately, the heat-treating equipment and the new product processing equipment were already half installed. Both of these areas had become "monuments" that could not be moved.

The quantitative study showed that the company had made a nonfixable blunder of relatively large proportions, particularly if they planned to use fork trucks for materials transport (as they had in another plant). Overlooked was the relative size differentials between the old and new forgings. The standard handling containers (tippable gondolas), in which they had a sizable investment, could only hold half as many new forgings as old forgings. As a result, for the same level of production, the new forgings required *twice* the number of fork truck trips as the old forgings. To make matters worse, in the company's 5-year projections, new forging production was expected to grow significantly with old production forecasted to decline. Using the straight-materials-flow approach (just following the lines on an operations process chart), the company had not taken into account either the current or future *intensity* of material flow or the materials handling costs associated with that intensity. Had they done so, the heat-treating department would have been located much closer to the new forging processing area

and much farther from the old forging processing area. Figure 7-1 (modified to ensure client confidentiality) and Table 7-1 show existing and alternative plant layouts and the differential annual cost savings that would have been achieved if an *engineered approach* had been used.

This is a classic example of what can happen when a subjective approach to plant layout is followed. Even though the straight-materials-flow method gives the *appearance* of being an objective approach, it can lead to serious trouble if intensity of flow is not taken into account. Also, the importance of relationships between activities not linked by material flow often are not given the attention they deserve. When this happens, one must be careful to avoid degenerating the planning process into a subjective template shuffling exercise until everyone is happy or it looks "about right."

The Consultant's Method or the Engineered Planning Approach

One of the other major problems associated with most of the above nonengineered methods is the tendency to produce far too many layout alternatives, particularly at the block layout stage. This is because the above approaches are, by definition, *subjective* in nature and difficult to measure. In my experience, if a company has developed more than four to six workable layout alternatives, it is a clear indication that an objective analysis *has not* been performed. It is sometimes an amusing experience to observe some "evaluation" meetings, in which layout teams have developed 15 or 20 layout drawings and taped them up on the walls of a conference room. Someone then poses the question to the group: "Well, which one do you think is best?" There is absolutely no way that someone can evaluate and differentiate between such a variety of layouts visually, without becoming totally confused and losing track of the objective.

When the straight-material-flow method is modified to take into account the intensity and cost of material flow, as well as nonflow-related affinities, much more efficient and effective layouts are attainable. We can first develop macro (block) layouts using a graphic approach similar to the method originally pioneered by Muther

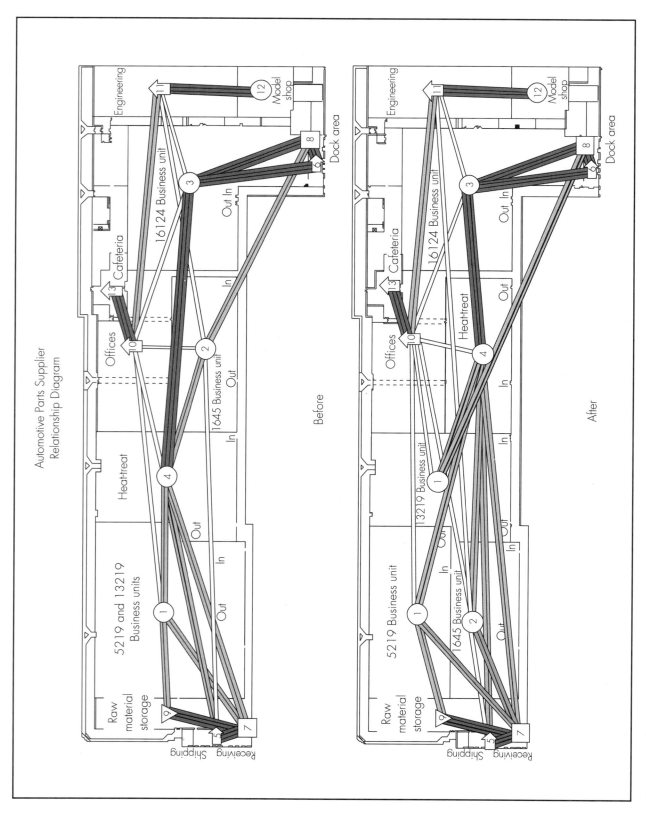

Figure 7-1. *An engineered approach to plant layout is the most effective insurance against subjective decision-making and its subsequent unnecessary cost.*

Table 7-1. Cost Savings Generated by an Engineered Approach to Plant Layout

	Distance—ft (m)	Cost	Time (min)
Existing facility	123,750,195 (37,719,594)	$604,622	806,989
Alternative facility	77,542,887 (23,635,071)	$425,969	567,834
Improvement	46,207,308 (14,083,987)	$178,653	239,156

in his work on systematic layout planning (SLP). After approvals are received for the macro layouts, we can develop more detailed layout alternatives (remember, there is always some overlap between the development of block layouts and detail layouts). When rough detailed layouts have been developed from the macro layouts we can quantify materials handling costs using computerized techniques. This will allow us to determine and evaluate several detail solution(s) for materials handling as well as the queue space allocations. This overall approach can generally be called *the consultant's method* or *the engineered planning approach*. (As a practicing consultant, I tend to be somewhat biased with regard to the name of the approach). Suffice it to say, most reputable consultants have to document their work *and* justify their recommendations. That is one of the major benefits of using the engineered approach. Consultants normally do not have the luxury of shuffling templates around on 15 or 20 layout drawings, taping them up on the wall, etc.

The engineered planning approach begins with macro layout techniques. These macro (or block) layouts can be done either manually or with the aid of specialized computer programs. We discuss computer-assisted methods later. We first need to discuss manual approaches, since most computer programs have their roots and algorithms in the manual methods.

THE THREE *As* OF PLANT LAYOUT

We discussed our overall layout planning approach in Chapter 3. The procedures are repeated in Figure 7-2.

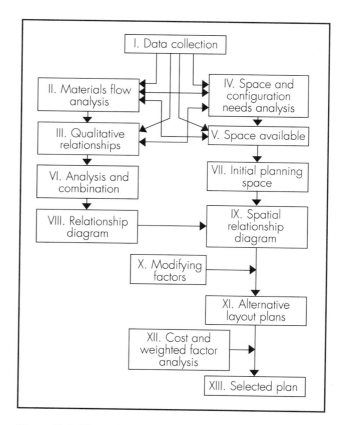

Figure 7-2. The major steps in the journey to a sound plant layout plan.

The overall procedure is divided into the three *As* of plant layout:

    Affinities (Relationships),

    Areas,

    Adjustment.

These three main task categories are shown overlaid on our procedure outline in Figure 7-3. *Every* facility layout, large or small, whether dealing with a factory floor, a master site plan, or the executive offices, involves the three *As*.

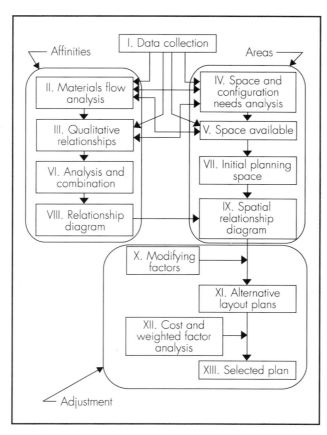

Figure 7-3. *The three As of plant layout—Affinities, Areas, and Adjustment—represent the task groupings of the overall layout procedure.*

METHODS OF ESTABLISHING RELATIONSHIPS BETWEEN ACTIVITIES

We begin our discussion by defining affinities and relationships between activities. For the newcomer to the facilities planning and plant layout field, it would help to show graphically the relationships between several activities. Figure 7-4 shows relationship lines drawn connecting several activities. The number of lines between the activity bubbles is proportional to the intensity of flow or the overall "closeness" desired

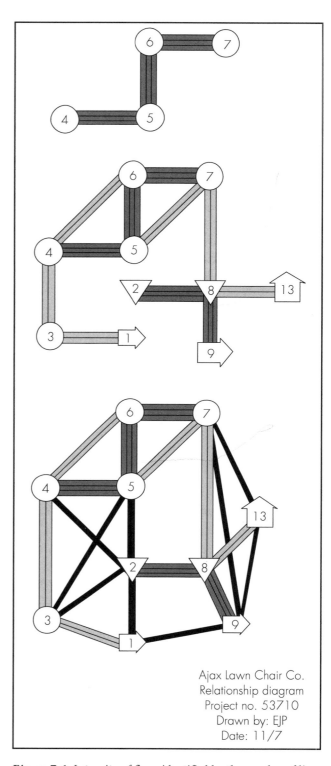

Ajax Lawn Chair Co.
Relationship diagram
Project no. 53710
Drawn by: EJP
Date: 11/7

Figure 7-4. *Intensity of flow, identified by the number of lines between activities on a relationship diagram, defines the desired closeness between activities.*

between the activities. The greater the number of lines, the more important that the two activities be close to one another. If the relationship between activity pairs has fewer lines, it does not have high "intensity" and the two activities can be proportionally farther apart. We discuss affinity/relationship bubble diagrams later.

Various authors, including myself, have used the terms affinities and relationships interchangeably. Several software-based layout programs have also intermixed the two terms. It would probably be more precise to define affinities as the quantitative and qualitative components of relationships. In other words, there can be two components to every relationship between activities, one quantitative and one qualitative. A relationship between two activities is really the combination of affinities between those two activities. A relationship is the *total* "tie" that determines how close an activity needs to be to another activity.

To avoid confusion, we refrain from using the term affinities, instead referring to all ties between activities solely as relationships (whether qualitative or quantitative).

Relationships can be determined quantitatively by way of material flow analyses and qualitatively using other techniques. This is collectively shown as procedures II and III in Figure 7-2.

Relationships Based on Materials Flow

In Chapter 4 we listed some of the detail data required during the data collection phase; this is block number I on our pattern of procedures shown in Figure 7-2. Now we need to move into material flow analysis—block number II on our pattern of procedures—an effort that requires a quantitative materials flow analysis. How do we quantify what is moving within the plant? We have already discussed some of this analysis. To *develop* the analysis, we will use our consolidated operations process chart and our from-to chart.

The consolidated operations process chart similar to the one shown in Figure 7-5 defines the interconnecting material *flowpaths* between activities. This chart is not meant to imply that these are necessarily the exact *physical paths* to be developed on our layout. All we can tell from

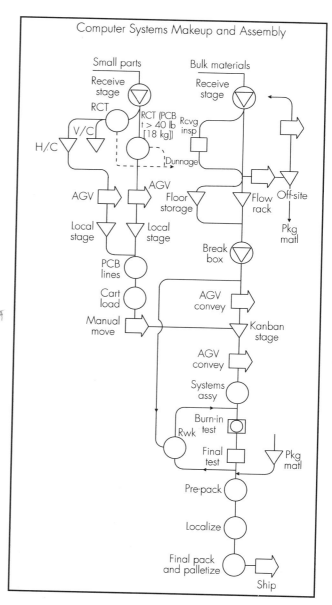

Figure 7-5. *Operations process charts such as this help the layout planner define the interconnecting materials flowpaths on a functional level, not necessarily the physical flowpaths themselves.*

the operations process chart is that there is some level of materials flow between the connected activities, but we have not as yet quantified that flow. You may remember that if we start to develop a layout directly from the operations process chart, without regard to intensity of the material flow, (the straight-material-flow method) we can sometimes get into trouble.

The missing link of the analysis is the intensity or strength of the materials flows between activities. We have the path connections. Where can we obtain the intensity levels of materials moving along those paths?—the from-to-chart.

Keep in mind that, for the purposes of developing macro layouts, we have made the very simplified assumption that intensity of flow is directly proportional to materials handling costs. That is not always correct, however; it often ignores, for example, the equipment and method selected for moving the materials. It assumes that all materials movement methods are homogeneous, (i.e., are equivalent unit loads). For instance, it assumes that two pairs of activities having the same intensity of flow between them would automatically have the same relationship between them and should, therefore, be the same distance apart. That would be an incorrect assumption if one path was served by a 100-ft (30-m)-long conveyor transport and the other path was served by a forklift truck driving 100 ft. It would be an even worse assumption if the forklift truck driver had to make a return trip empty to pick up each load.

The point to remember when dealing with materials flow relationships is that, in the final analysis, *optimizing distances between activities should be based on materials handling cost* between those activities (and not necessarily materials flow intensity).

Our starting problem for developing a macro layout is that we normally do not know what mode of materials handling equipment will be selected. We will need to adjust the simplifying intensity assumptions later when we are closer to developing detail layouts.

Complex (and expensive) custom-designed software programs can generate relationship charts and diagrams directly from the from-to chart input. However, unless the user is a professional layout planner who spends a great deal of time developing layout plans, custom-developed software is generally not justified.

The first thing we need to do after completing the from-to chart is develop a ranked bar chart similar to the one shown in Figure 7-6. This chart simply shows the equivalent unit loads of materials moving between each activity pair in descending, ranked order. This information is taken directly from each of the intersecting cells on our consolidated from-to chart.

It is a relatively simple procedure to manually sort the ranked materials flows if a computer-generated spreadsheet is not used. All that is required is a completed from-to chart to move directly to a ranked bar chart like that shown in Figure 7-6.

The flow of materials intensities shown as numerical weighted equivalents on the from-to chart should then be converted to the conventional vowel/letter designations (shown in Figure 7-7) used for recording the information into a classic activity relationship chart. These vowel/letter conventions have become facilities planning standards over the last 40 or 50 years. To prevent confusion, it is important to use them for the duration of the layout project. The intensity designations are:

A = **A**bnormally high intensity of flow
E = **E**specially high intensity of flow
I = **I**mportant intensity of flow
O = **O**rdinary intensity of flow
U = **U**nimportant moves of negligible intensity

We will use these designations to assign ratings to each of the material flow intensity bars on our chart. When numerical equivalents are used on a from-to chart, the total intensity of movement from and to each department can be calculated by adding the from and to values for the particular activity. As mentioned in Chapter 5, entries above the diagonal line represent linear forward movement, while those below represent countermovement or backflows.

Steps in the procedure for developing the ranked materials flow (equivalent unit) bar chart:

1. Identify each route or activity pair between which materials move.
2. Record the quantitative level of materials moves on a from-to chart as outlined earlier.
3. Rank the flows from highest to lowest.
4. Prepare a bar chart similar to the one shown in Figure 7-6.
5. Using logical breakoff points, divide the chart into groupings of flows ranked by the vowel/letter rating system. Keep in mind that according to the 80/20 rule, the *A* pairs may only comprise a small percentage of the flows. There will be more *E*s than *A*s, and still more *I* activity pairs, and so on.

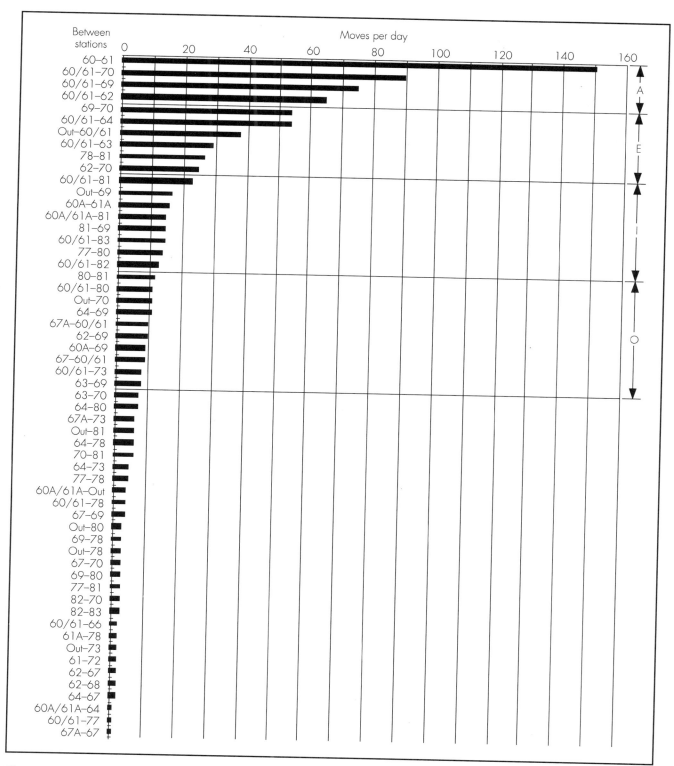

Figure 7-6. *By ranking and plotting materials flow frequency, the planner can easily visualize activity flow intensities in prepara-tion for creating an activity relationship chart.*

Vowel/Letter Conventions for Diagramming Materials Flow Intensities

Vowel letter	Number value	Number of lines	Material flow intensity	Color code
A	4	////	Abnormally high flow	Red
E	3	///	Especially high flow	Orange yellow
I	2	//	Important flow	Green
O	1	/	Ordinary flow	Blue
U	O		Unimportant flow	—

Vowel/Letter Conventions for Diagramming Relationships

Vowel letter	Number value	Number of lines	Requirements for areas to be close	Color code
A	4	////	Absolutely necessary	Red
E	3	///	Especially necessary	Orange yellow
I	2	//	Important	Green
O	1	/	Ordinary closeness OK	Blue
U	O		Unimportant	—
X	$-1, -2, -3, ?$	$\wedge\wedge\wedge\wedge$	Not desirable	Brown

Figure 7-7. Vowel/letter designations used to define materials flow intensities are really predictors of the closeness required between activity pairs.

The resultant rankings define the desired closeness relationships between activities based solely on materials flow. In a perfect world, with the same mode of materials handling equipment used for all moves, (e.g., forklift truck, push cart) the activity pairs and ranked bars would have the exact same descending rank order if we were measuring and ranking materials handling costs. Instead of the bars showing equivalent unit loads moving between activities, the bars could be calibrated in terms of materials handling monetary cost. We would want to optimize our layout so the activity pairs shown at the top of the chart would be very close together in our final layout.

In other words, because the two activities defined by the top bar in the chart would have the highest number of equivalent movements (ranked with an *A* designator), they should be located as near as possible to one another in our plant layout. If we place them far apart, total materials handling cost will go up. It is far better to have the activity pairs at the top of the chart closer to each other than the activity pairs at the bottom of the chart. The ranking then actually defines the desired closeness rating between each activity pair. Before finalizing relationships between activities, we need to take this analysis a bit further, since there may be other factors affecting closeness that are unrelated to materials flow.

Relationships Based on Nonflow Factors

A variation of the vowel/letter-based evaluation system is used as a basis for charting nonflow relationship input to the layout process. This data is usually collected through interviews with management, line supervisors, marketing personnel, shop support workers, receiving and shipping personnel, and others. Each avenue of input should be considered; however, all inputs cannot be considered equal; some may need to be weighted.

There are several reasons why materials flow relationships alone cannot be used as the sole basis for layout design.

- Since supporting services integrate with flow, the location of these services in relation to the services they perform must be considered.

- In some industries (such as jewelry and some types of electronics manufacturing) materials flow is relatively unimportant, since only a few ounces (grams) or pounds (kilograms) of materials are moved in any day (more packaging materials may be moved than actual product). In other process-dominated industries, materials may be piped directly from process to process and flow may not be the most important factor.

- In service industries, office areas, or maintenance and repair shops, often no real or definitive flow of materials exists. That being the case, some logical method beyond materials flow analysis must be used in relating areas within these industries.

- Even in industries involving heavy materials flow, which is often the basis for process sequence and equipment arrangements, flow of materials is only one reason for placing operations tangential to one another. Flow reasons may conflict with or require adjustments in other requirements such as safety. While a particular routing may suggest that the sequence be form, trim, treat, subassemble, assemble, and pack, the treating process may be both messy and dangerous, suggesting placement of that activity elsewhere in the flow.

For these reasons, the intensity of materials flow may be overridden in arranging the layout. The distribution of utilities, the cost of quality control, and product contamination also can be legitimate reasons to use factors other than flow in assigning proximity relationships.

Usually, there are at least half a dozen "nonproduction" areas where there is no materials flow. Nonflow relationships have to be established from there to other areas. Typical of these areas are:

- Maintenance,
- Tool storage, & fixtures
- Lunchrooms,
- Restrooms,
- Quality assurance,
- General offices,
- Locker rooms.

Experienced layout planners use a variety of methods to establish nonflow relationships, a few of which follow.

- A highly experienced individual within the plant establishes his own rankings based on his personal knowledge of both materials flow and service relationships. In effect, he develops one relationship chart with the combining of relationships done manually.
- Since many layouts are based on producing cost reductions in the materials handling effort, relationship rankings based on materials flow only can sometimes be used initially. If some service relationships are indeed ranked high, they are inserted at a later time during the special relationship phase or during the drawing of the actual block layout.
- Interviews are held with key operations and management personnel. Their input and agreement is received *prior* to the development of an overall relationship diagram. It is imperative that an agreement is reached in this abstract analysis phase before any lines are put on paper or in the CAD system.
- An experienced internal employee who is divorced from the organizational politics, or an outside consultant, is employed to develop the flows, relationships, relationship diagrams, etc.

However the relationships are established, it is important that factory supervision and management people are included in the approval stage. They must be made to feel that they are part of the team, especially those that are responsible for materials handling and operations after the layout is implemented.

Similar to a folded from-to chart, the cross-sectional mileage-type chart is one means of recording the relationship of areas that incorporate nonflow-dominated criteria into the decision process (see Figure 7-8). This classic relationship chart has become a standard for recording relationships. It incorporates not only closeness criteria, but also avoidance criteria. It integrates support, office, and service areas with production areas to help provide a logic pattern for our layout.

Each diamond-shaped box integrates information about the activity pair it represents. Logically, it is like a from-to chart folded diagonally so that the from-to and the to-from boxes fall on top of each other. Each diamond-shaped box shows the total relationship in both directions between two activities.

The upper half of each box shows the importance ranking of the relationship; the lower half shows the code for the reason or reasons for this importance ranking.

To use the chart, read from the left, starting with the numbered column listing all activities. Then evaluate the nonflow relationship of each activity to every other activity using the following six categories, which are listed in descending order of importance:

- **Functions designated by *A*, that *absolutely must be placed close together*.** These functions must be close because the interface or communication between them is very high. The relationship between the incoming inspection department and the receiving docks may need to be an *A* if there is a high level of interaction. Occasionally, functions might need to be very close for security reasons. Placing a shower immediately adjacent to an area containing acids that must be handled during production is within this category for legal as well as safety reasons. A receiving dock may need to be placed immediately next to the shipping dock so that personnel and equipment can be shared. We would either combine these two areas into one activity or we could establish an *A* relationship between them (even though there may be zero direct materials flow between them). In public service businesses, client or customer convenience and image may become the primary reasons for absolute closeness.
- **Functions designated by *E*, that would be *especially beneficial* if placed close together.** These functions can exist without being tangential, but many benefits will accrue from placing them as close as possible. Putting the receiving area close to the materials storage area often fits into this category. So does placing a private secretary adjacent to his or her boss in a law office. This last relationship may be less important in other businesses.
- **Functions designated by *I*, that are *important* to be placed close together.** These functions will not be radically affected by some distance between them, but the operation will

run smoother if they are not too far apart. The activity pairs do not have to be as close as an *A* or *E* relationship pair. It may be important to have a high-tech manufacturing installation close to the visitor's entrance for image purposes. In an office environment, it is important that the manager's clerical staff be close to the manager's office because of the closeness with which these functions must work. But, placing these functions 50 ft (15 m) down the hallway would seldom be as detrimental to plant operations as separating critical production functions.

- **Functions designated by *O*, whose *ordinary closeness* is acceptable.** Placing these functions farther apart will not prove as costly as separating functions in the first three categories. They may be traditionally close together, and should remain near each other for reasons of morale, unless such placement interferes with the placement of functions listed in the first three categories. An example would be the relationship between employee locker rooms and the lunchroom. While filing areas are traditionally close to clerical pools in an office environment, and should remain close if possible, separating these functions is possible without affecting operations. A pneumatic tube system can carry file copy relatively long distances if necessary.
- **Functions designated by *U*, whose locations relative to each other are *unimportant*.** These functions can go almost any place as long as space is provided. This category could include such things as storage areas for inactive files or for some safety stock seldom needed. The safety stock is kept on the premises because breakdown would delay production several months while a replacement is manufactured.
- **Functions designated by *X*, which should *never be together*.** (Unless a suitable physical barrier can be erected between the activity pair.) These functions include storage areas for chemicals that can explode if they accidentally come into contact with each other, solvent storage and welding, paint areas which can cause contamination of some production or office functions, and any areas which must legally be separated.

While these *A*, *E*, *I*, *O*, *U*, and *X* relationship classifications are used for establishing closeness on all manufacturing layouts, the reasons for each classification will usually vary with each layout project. It is important that these reasons be listed on the relationship chart. Some of the more common reasons for rating supporting relationships are:

- Flow of support materials,
- Equipment sharing,
- Personnel sharing,
- Management control,
- Floor loading,
- Personal contact/communication,
- Management directive,
- Security,
- Noise/dirt,
- High-bay plant area required,
- Safety.

Two or three reasons may be marked in the lower half of a single diamond block when multiple reasons for a ranking exist. While a specific ranking may be based on a number of reasons, most layout projects require no more than eight or ten reasons. More can be used if necessary, of course.

A completed classical relationship chart is shown in Figure 7-9. It includes general information on ratings and reasons. *O* and *U* relationships are not shown (for the sake of clarity).

Color coding

Because a great deal of information is compiled on a single chart, many planners find it easier to gather meaning if the classifications are color coded. The following colors have become conventional for this type of chart.

A classifications	red
E classifications	yellow
I classifications	green
O classifications	blue
U classifications	uncolored
X classifications	brown

Often, color coding is added after the relationships have been set to avoid having to change color categories. In this case, the center of the triangle remains uncolored, while the triangle is rimmed with a band of color.

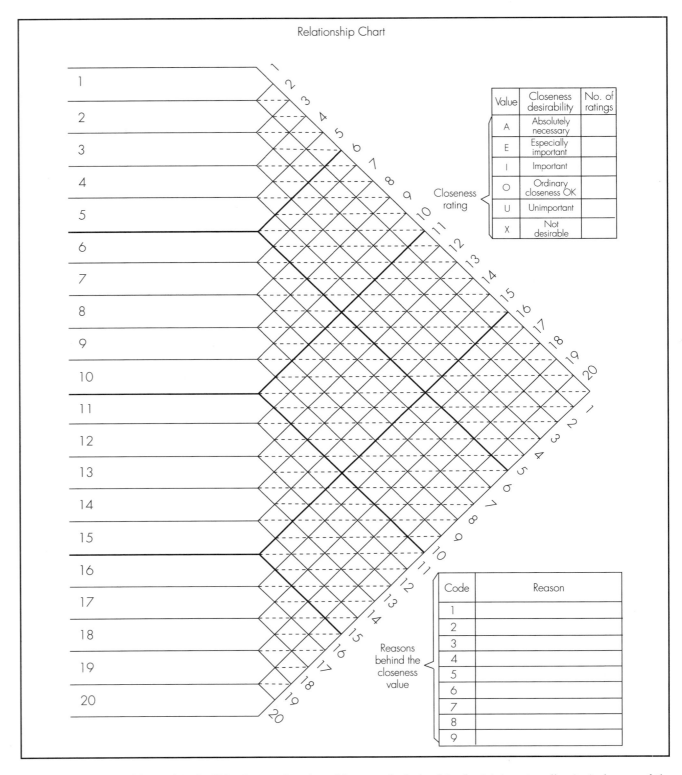

Figure 7-8. *Most widely used in facilities layout planning, this type of relationship chart integrates all principal areas of the enterprise—support, office, service, and production—and provides closeness and avoidance criteria.*

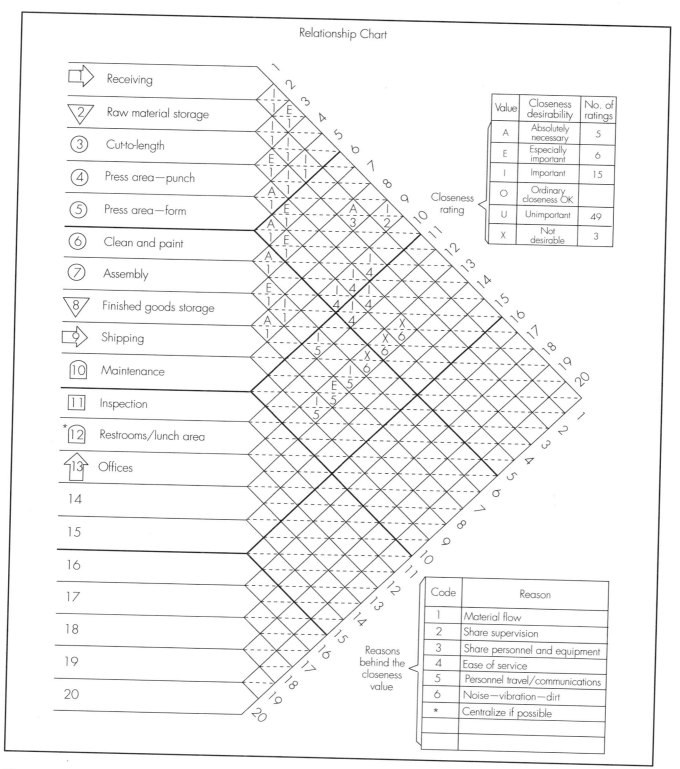

Figure 7-9. A completed relationship chart gives a single-point view of activity pair relationship ratings and reasons.

Refining classifications procedures

The number of ranks is limited. The number of relationships assigned to each also should be limited. Those who are not used to this charting procedure often overassign *A* classifications. A good rule of thumb is to limit the percentages of total possible classifications to:

A   2% to 5%
E   3% to 10%
I   5% to 15%
O   10% to 25%
U   The remaining functions.

For example, if we have 20 activities or work cells that must be laid out in relation to one another, we would have a potential total of 190 relationship pairs.

$$\frac{n\,(n-1)}{2} = \frac{20\,(20-1)}{2} = 190$$

If we were perfect layout planners, we would have approximately eight *A* relationships (190 × 4%), 13 *E* relationships (190 × 7%), and so on.

In cases where *X* classifications are really based on highly dangerous situations, an *XX* classification may be added, colored black. In actual practice, heavy walls, technical solutions, or other physical barriers are constructed between activities with *XX* relationships.

The charting process

The relationship chart may be used to (a) categorize only those nonflow relationships, (b) to show only flow relationships, or (c) to show *combined flow* and *other-than-flow* relationships.

When charting flow and nonflow relationships separately,

• Develop the intensities of flow for each operating activity to every other activity;
• Rank the intensities between each pair of activities using the *A, E, I, O, U, X* system;
• Color code the chart.

On a separate chart:

• Develop the rankings and reasons for relationships among nonflow activities;
• Rank the relationship between each pair of activities using the *A, E, I, O, U, X* system;
• Color code the chart.

Combining the resultant two relationship charts is recommended in 60 to 70% of all projects. In those remaining projects, either flow or nonflow factors will be so obviously dominant that the uncharted factors will not alter the layout pattern.

Figure 7-10 shows where we are at this point in our analyses. Now we need to *combine* flow and nonflow relationships to get one overall relationship chart.

Combining and Weighting Flow and Nonflow Relationships

A weighting scheme or judgmental scheme is used by some planners and academics to combine flow and nonflow relationships to form one combined chart. The start of this combining process can usually be set up on a microcomputer-based spreadsheet format. In my experience with manufacturing plants, the most prevalent weighting scheme will weight material flow relationships as twice as important as nonflow relationships. Generally, weighting factors above a value of 2.5 are not used, since higher values may tend to block out and completely overshadow important nonflow relationships. (The vowel/letter scoring conventions are shown in Figure 7-7.)

The matrix in Figure 7-11 illustrates one way to combine relationships using the conventional scoring techniques. This particular combination sheet weights materials flow rankings as twice as important as nonflow rankings. The first column lists all of the activity pairs. The second column shows the materials flow intensity rankings assigned to the activity pair. The third column shows the conventional scores for each ranking (e.g., *A* = 4, *E* = 3, *I* = 2 and so on). The fourth column shows all materials flow rankings multiplied by the weighting factor of two. Similar to the materials flow columns, the fifth and sixth columns show the nonflow derived rankings and their respective scores. The seventh column shows the additions (combinations) of the scores in the fourth and sixth columns for each activity pair. This column contains the combined scores. Our next step would be to take these new, combined scores and develop a new ranked bar chart. We can then reapply the

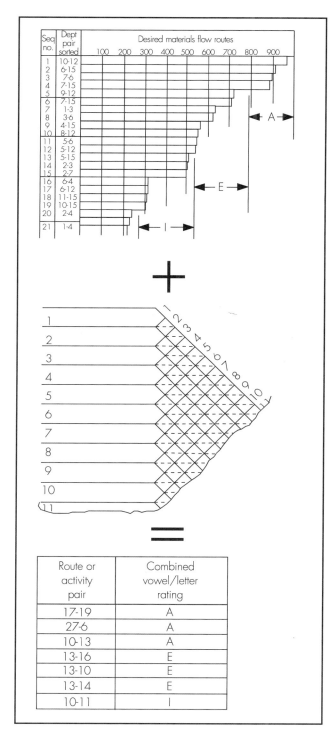

Figure 7-10. *At this stage of the layout process, the layout planners should have a relationship chart that combines all of the rankings of both the materials flow analysis and the service/support relationships.*

vowel/letter ranking system to get the final combined and ranked data. This data is then transferred to a standard relationship chart, and the reasons for the relationships are written in the bottom of the appropriate diamond blocks.

The planner must be careful during the combination process that his or her weighting scheme does not completely overshadow very important nonflow relationships. For example, I have found that it is far better to do separate relationship analyses for production operations and office blocks of space. When one tries to mix the two in one combined analysis, suboptimum results are obtained. Relationship weighting schemes tend to incorrectly lower the relationship values between those activity pairs where no materials flow exists but where strong nonflow-based relationships do exist.

For instance, Figure 7-12 shows an excerpt from a typical weighting scheme. The chart in the upper left-hand portion of the figure shows the relationships between activities based purely on materials flow. The abbreviated chart in the upper right-hand portion of the figure shows the relationships between activities based purely on nonflow considerations. Only the materials flow ratings (using standard point scores) are weighted upwards by multiplying by a factor of two. This is shown in the upper middle portion of the figure. Note that there is no materials flow between activities 4 and 6. However, there is a very strong relationship (*A*) between these two activities based on nonflow considerations. Note also what happens when we combine the flow and nonflow ratings at the bottom of Figure 7-12: activity pair 4-6 *has now been downgraded* to an *I* relationship. Is this what the planner really wants? Initially the team felt that a very strong nonflow *A* relationship was required between these activities. Some of that initial importance was inadvertently lost in the combination process. This is the major pitfall of most weighting schemes. Computer-based weighting schemes can exacerbate the problem if the output is taken for granted as being correct without a detailed inspection of the results. Most combination schemes, whether manual or computer-driven, require a manual correction or adjustment to some relationships.

Combining Relationships

Activity pairs	Materials flow intensity	Conventional scores	Weight factors (2)	Nonflow ratings	Nonflow scores	Combined scores
2-5	E	3	6	—	—	6
3-11	O	1	2	E	3	5
5-7	U	—	—	—	—	—
7-9	A	4	8	I	2	10
2-4	A	4	8	E	3	11
3-4	E	3	6	O	1	7
4-9	I	2	4	—	—	4
2-3	I	2	4	A	4	8

Combined Chart

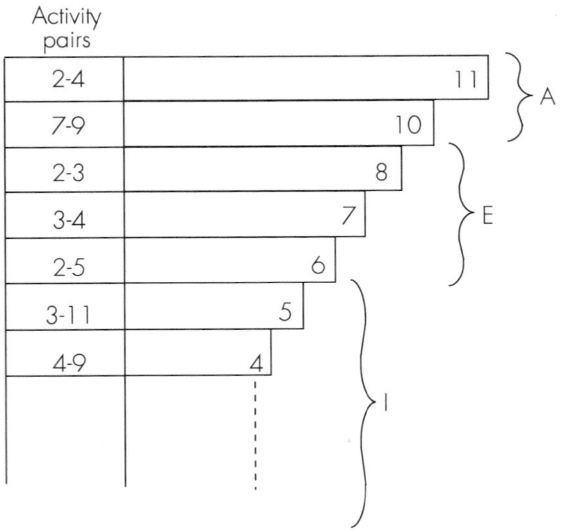

Activity pairs

2-4	11	} A
7-9	10	
2-3	8	} E
3-4	7	
2-5	6	
3-11	5	
4-9	4	} I

Figure 7-11. *Combining relationships using conventional scoring techniques. Combined scores derived from this type of chart are used to develop a new ranked bar chart.*

These are the principal reasons that most relationship analyses are done off-line. That is to say, relationship combinations are best done manually and not with a computer. For example, if there is an *A* relationship between shipping and receiving and we want to keep that strong tie, we have to *manually* review the combined relationships and make adjustments where needed (this is not a difficult process). We can certainly use a computer up to the point of

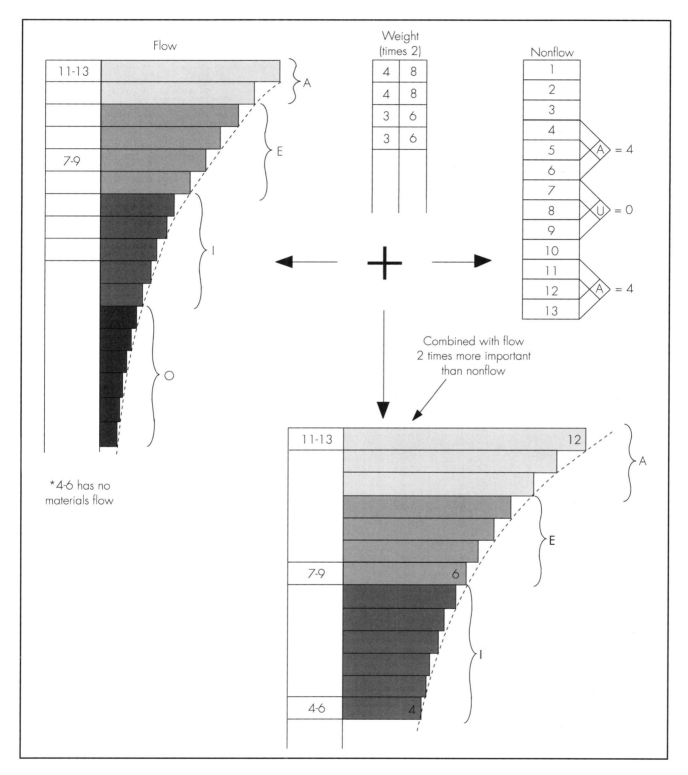

Figure 7-12. *Care must be taken during the combination process to avoid generating incorrect relationship values. Best practice argues in favor of conducting separate analyses for materials flow and nonflow relationships.*

relationship combination, and we can make very good use of a computer after the combination. But, at the time of this writing, dozens, if not hundreds, of researchers have tried for more than a quarter of a century to develop algorithms to handle relationship combinations without success. Most of the usable computer programs available today still rely on the planner to combine flow and nonflow relationships, prior to the computer developing block layouts.

A simple alternative approach in helping to avoid these problems is to only perform the combination exercise on those relationships that have a materials flow component. Those relationships based solely on nonflow considerations will keep their starting relationship scores. (A manual review of all relationships will still be required after combination.)

After (and even before) the development of the combined relationships, two different schools of thought come into play. When early versions of this type of analysis were originally put in use back in the 1960s, pure academics and some consultants of that era felt that the relationship diagram or activity/proximity diagrams should be drawn from the combined relationship chart without regard to space requirements. With the advent of computer-based analysis, that original view has been modified. However, we continue with the classical approach here since it is more easily learned than trying to integrate with space needs immediately.

It is important to recognize that, in the industrial real world, there is an ongoing integration (and ongoing revisions) of space configuration and relationships from the very start of the analysis. One facet or the other cannot be considered in abstract without keeping both in mind. As the pattern of procedures indicates in Figure 7-2, heavy interaction exists among several tasks at the start of the analysis. There also may be continuing interaction during other phases of the analysis.

The attempt to keep space needs separate from relationship development is the theoretician's effort to develop a true or *pure* relationship diagram. One must be careful not to fall into this trap in actual practice. "Ideal" layouts can be given up rather quickly when a piece of equipment does not fit into the space allotted for it!

Other Important Relationships (Affinities)

From our discussion of relationships between activities based on materials flow and nonflow factors, the planner should recognize that there may be several other building or site-specific, nonflow factors, that can affect the plant layout.

Many ideal plant layouts are implemented and work very well for a year or so, until production increases or new products are added and the company discovers it needs to expand. Frequently, many companies also discover that their ideal layout has boxed in a department that needs to be expanded. To rearrange the plant properly, it's quite possible that several functional areas and many pieces of equipment will need to be moved at a relatively high cost to make room for the department or manufacturing cell that needs to expand. Each move in "domino effect" affects other areas of the plant that also may have to move or be rearranged.

In one particular project, I was engaged to help a company after they had moved their welding department (and other work centers) three times in 3 1/2 years. Some 4 1/2 years before, a graduate student at a local university helped the company develop an ideal layout using relationships similar to those discussed here. The layout worked fine for almost a year, but soon turned into a disaster in the plant manager's eyes. What both the company management team and the student had neglected to account for was *flexibility*.

The lack of flexibility in plant layouts is one of the most serious problems a company can face. Gradually over time the problem can add myriad indirect costs that are difficult to isolate and quantify. But there is a method that promises to give us at least some assistance in accommodating the *next* expansion. It involves an *expansion* or *bottleneck sensitivity analysis*.

Just as we ranked materials flows, we should also rank activities by the *probability of needing the next area expansion*. This normally involves a capacity analysis for all of the activity areas (which most companies should already have and should update regularly). Figure 7-13 shows a partial activity listing for an example manufacturing plant. The bottleneck or choke function with the highest probability of needing more

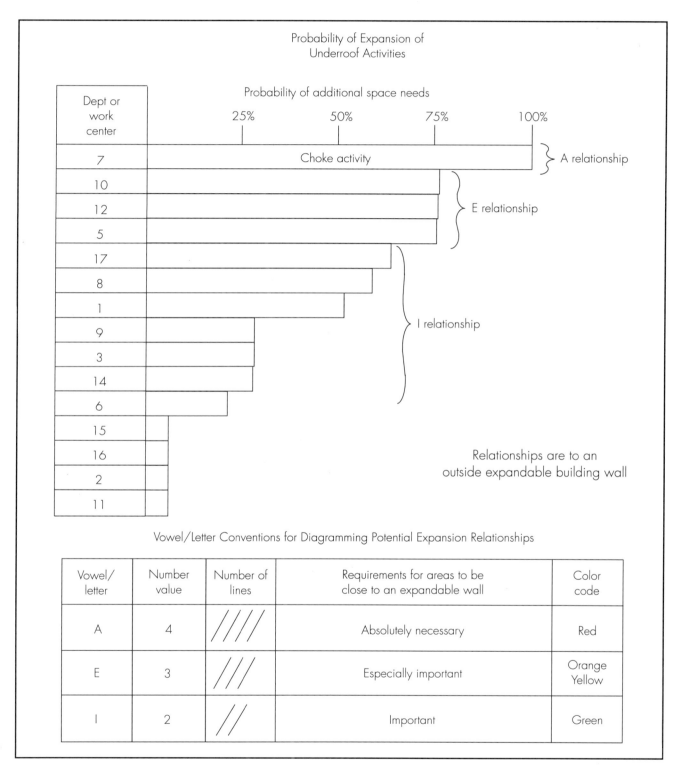

Figure 7-13. *In a future expansion analysis, the bar chart and vowel/letter designations identify the urgency of expansion space needs by department or work center.*

space is listed at the top of the bar chart. The activity with the next highest probability for expansion is listed next, and so on.

What are we going to do with this probability chart? We are going to *define relationships* between all of these activities and an expandable portion of the building. The top bottleneck operation in our bar chart should have at least an *A+* relationship between it and a wall on an expandable side of our building. Similar to the bar chart used to rank our materials flow intensities, this new chart can use the vowel/letter ranking system to rank expansion probability intensity. In practice, it is more likely that the planner will start with relationships and actually fix the location for the bottleneck function similar to fixing the location of a monument. It is usually a yes-or-no decision and not subject to "grades" of closeness to an expandable portion of the facility. The planner also should realize that we have simplified the bottleneck analysis by assuming static probabilities. In an industrial setting, a dynamic computer simulation of bottlenecks may be required, since bottlenecks will constantly change and shift as we make improvements to the production processes.

Some activity pairs may involve relationships of various activities to items or locations, such as:

- Natural sunlight (windows, skylights, etc.);
- Dock doors to a compass direction (not facing north in northern latitudes, prevailing winds, etc.);
- Magnetic north (has relevance in some high-tech testing areas);
- Parking areas, employee entrances, etc.;
- Customer entrance.

There are many more physical facility constraints that also could have fixed relationships to various activities: areas of the plant that can withstand heavy floor loadings, for instance; or building clear heights, utilities supply points, rail and highway access, etc.

GETTING THE LOGIC APPROVALS

In the interest of clarity, from this point (unless specified otherwise), the term relationship(s) refers to the total combined relationship(s) that take into account both flow and nonflow considerations.

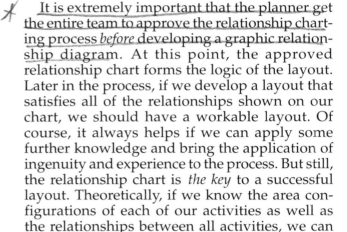

It is extremely important that the planner get the entire team to approve the relationship charting process *before* developing a graphic relationship diagram. At this point, the approved relationship chart forms the logic of the layout. Later in the process, if we develop a layout that satisfies all of the relationships shown on our chart, we should have a workable layout. Of course, it always helps if we can apply some further knowledge and bring the application of ingenuity and experience to the process. But still, the relationship chart is *the key* to a successful layout. Theoretically, if we know the area configurations of each of our activities as well as the relationships between all activities, we can input that data into a computer and have the layout generated automatically. (In practice it is not quite that easy, but several computer programs are available that may help in that regard.)

A good layout will show all activity pairs, with an *A* relationship between them, very close together, if not abutting. Those activities with *E* relationships between them would be farther apart. They would still be close to one another, but not as close as the activity pairs with *A* relationships. Those activity pairs with *I* relationships between them would be somewhat farther apart than those with *E* relationships, and so on. In other words, closeness of two activities is proportional to the strength of the relationship between them.

The relationship chart, coupled with all of the reason codes assigned, forms the very foundation or logic basis for our macro layout planning work. Therefore, we need to be sure that all involved with the planning process agree to the relationships that have been established. *This agreement and approval process must be completed first, before anyone sees alternative layout plans.* The approvers should not be distracted with new layout proposals prior to approving the relationship chart. If working on the rearrangement or expansion of an existing plant, it is quite acceptable (and probably desirable) to have an existing plant layout drawing to refer to when reviewing the relationship chart. However, if others have preconceived new layouts that have not been developed using the materials flow and relationship analysis discussed here, the approval process may bog down or, worse yet, become biased. It is far better to review any newly proposed layouts *after* everyone agrees to the relationship analysis.

When working on relationship charts for macro layouts, the planner is usually in the block layout stage. Although work center teams may be tangentially involved, the logic (relationships) approvals should come from individuals who are at higher levels in the organization. The work center teams will become more involved later, when the confines of their particular operations have been determined. The approvers are typically middle management and sometimes senior management personnel. The approval team should definitely include whomever has the responsibility for approving the subsequent layout alternatives. These individuals also should be generally familiar with how much or how little the various functions and departments interact with each other.

Again, it should be stressed that the approved relationship chart forms the foundation or "measuring stick" for much of our subsequent plant layout work. If we do a good job in determining relationships, we will generally be able to quickly develop good plant layout alternatives. Conversely, if we do a poor job in defining relationships, we will develop poor plant layout alternatives. How well the subsequent layout alternatives match the original relationships assigned can be both qualitatively and quantitatively measured.

You should not be left with the impression that relationship charts, once approved, are inviolate and can never be changed. On the contrary, on complex layouts, it is easy to overlook one or two important relationships. These "misses" show up when we have developed several block layout alternatives and they are in the process of being reviewed. At that point, however, so much of the other relationship logic will have been met that it normally takes only a minor amount of revision to accommodate the changes. The more important aspect is that all of the people involved in evaluating the block layouts will have been part of the team who developed the original relationships. They will feel a sense of ownership and a certain amount of pride in the results. By using this objective relationship analysis process, it is usually easy

to convert naysayers (who juggle templates around till it "looks about right") into active, contributing participants.

Getting everyone involved to approve the relationship analysis, however, is not an easy task. It usually requires educating management-level people who may have never heard of relationships being used in an industrial setting. If at all possible, it generally pays to have the managers take at least a half day and preferably a full day of training in plant layout before starting the process.

THE AFFINITY, PROXIMITY, RELATIONSHIP, BUBBLE DIAGRAM— TAKE YOUR CHOICE

The title of this section includes just four of the names that have been used to describe the diagrams that we will construct from the data in our relationship chart. To be consistent, we use the term *relationship diagram* throughout this text.

We should now have thoroughly completed our flow-of-materials work and our supporting services relationship work. These have been combined in previous sections of the book into a combined relationship chart. We now need to begin our graphical or "visualization" process by diagramming the relationships. We now have the relative closeness importance of each activity to every other activity in our relationship chart listing.

It is usually a help to rank the relationship chart listing in descending order of importance, as shown in Figure 7-14. An acceptable relationship diagram of that ranked listing must now be established.

Bubbles or symbols are drawn for each activity that is represented in our layout. Keep in mind that we have not as yet discussed space requirements. Some planners use the same size bubble or symbol for each activity. However, if we do know the space requirements or space estimates for each activity at this time, it is better to make the size of each bubble equivalent to, or roughly equivalent to, the proportional size of the area needed, similar to that shown in Figure 7-15. This is a common practice used by nearly all architects on space planning projects.

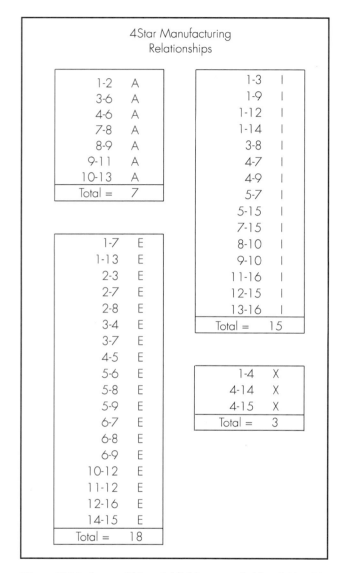

4Star Manufacturing
Relationships

1-2	A
3-6	A
4-6	A
7-8	A
8-9	A
9-11	A
10-13	A
Total =	7

1-3	I
1-9	I
1-12	I
1-14	I
3-8	I
4-7	I
4-9	I
5-7	I
5-15	I
7-15	I
8-10	I
9-10	I
11-16	I
12-15	I
13-16	I
Total =	15

1-7	E
1-13	E
2-3	E
2-7	E
2-8	E
3-4	E
3-7	E
4-5	E
5-6	E
5-8	E
5-9	E
6-7	E
6-8	E
6-9	E
10-12	E
11-12	E
12-16	E
14-15	E
Total =	18

1-4	X
4-14	X
4-15	X
Total =	3

Figure 7-14. As an aid to establishing a workable relationship diagram, it is important to first list the activity relationships by vowel/letter groups.

For purposes of training individuals who may have never done plant layouts, we assume that we have not as yet determined our space needs and are working purely with area symbols or the same size bubbles.

The planner should start his or her diagram with the *A* relationships first. The other relationships are charted in a similar fashion starting with the *Es*, then the *Is*, and so on. It is customary to use lines or bars to indicate flow or rela-

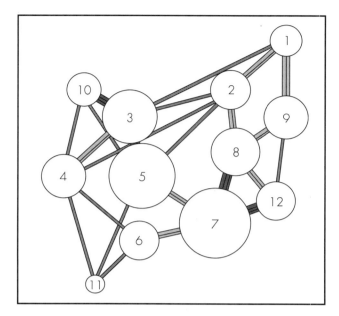

Figure 7-15. Bubble diagrams, with the size of the "bubbles" representing the proportional size of the area needed, graphically depict materials flow intensities and relative space requirements of each activity.

tive importance even though we may be working with nonflow relationships. Figure 7-16 shows the line conventions normally used for connecting or diagramming the relationships. Four lines or three bars signify an *A* relationship, three lines or two bars an *E* relationship, and so on.

It should be stressed that diagramming relationships manually (and sometimes even with a computer as an aid) is an iterative process. It is not unusual to go through four, five, or even six or more trials and refinements before an acceptable diagram is achieved. That is why a computer routine can be so useful in this phase of the work. Keep in mind, however, that most diagrams continue to be fashioned manually, even though a graphics-oriented computer and specialized programs can be of great help.

The method of establishing the diagram itself consists of connecting the activities by lines or bars whose lengths and widths represent the original relationships charted. The number of lines or bars (width) represents the intensity of the relationship or the degree of closeness desired. The lengths of the lines or bars determine

			Vowel/Letter Conventions for Diagramming Relationships		
Vowel letter	Number value	Number of lines	Requirements for areas to be close	Color code
A	4	////	Absolutely necessary	Red
E	3	///	Especially necessary	Orange yellow
I	2	//	Important	Green
O	1	/	Ordinary closeness OK	Blue
U	O		Unimportant	—
X	−1,−2, −3, ?	VVVVV	Not desirable	Brown

Figure 7-16. Critical to the clarity of relationship diagrams are the established conventions denoting "closeness" levels.

how well we meet these tests of relationships. In a perfectly drawn diagram, all of the *A* relationship connections would be the same length and also the shortest length of any other relationships drawn; all the *E* relationship connections would be the same length—perhaps 1 1/2 to 2 times as long as the *As*, but still shorter than any other lower value relationships; all the *I* relationship connections would be the same length—perhaps 1 1/2 to 2 times as long as the *Es*, but still shorter than any other lower value relationships, etc. It helps to think of the graphic lines of the relationships as rubber bands pulling activities together. The more lines, the more rubber bands, and the greater the strength or pull of the connecting relationship, as shown in Figure 7-17.

A dashed line or a sine wave-type line is normally used to diagram the undesirable relationships or *X* values. It is sometimes helpful to think of these as compression springs pushing the activities apart.

The planner should work from the simple to the complex. Frequently, the ordinary *O* relationships are not even shown on a diagram with many activities. That is one of the real benefits of using a specialized computer program for generating relationship diagrams; a good program takes into account *all* relationships and adjusts, or optimizes, the diagram accordingly.

Sometimes it is wise to start the diagram with a hub department or activity. These activities stand out as the ones with the most activity relationship connections. Whenever a supporting service activity has many relationship connections, it is wise to consider decentralizing such an activity. This is typical in the case of commonly used locations, such as storerooms, restrooms, etc. See Figures 7-18 and 7-19 for example diagrams. In Figure 7-18, departments no. 5 and 22 may be considered hub activities. In Figure 7-19 the service computer access has been eliminated by using distributed terminals/micros at individual workstations.

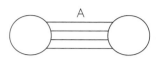

Strongest pull or affinity between two activities—
requires the activities be as close as possible in
our layout (absolutely necessary to be close)

Next strongest pull or affinity between two
activities—not as strong a pull as an *A*, but still
very strong (especially important to be close)

Important pull or affinity between two activities—
not as strong as an *E* and much less strong than an *A*,
but still a strong affinity

Activity pairs with *A* relationships should be drawn
very close to one another, as shown.

Activity pairs with *E* relationships should be drawn
close to one another but not as close as *A* pairs.
Generally try to keep *Es* about 1.5 times farther
apart than *As* in the preliminary diagramming stage.

Generally try to keep *Is* about 2 times farther apart
than *Es* in the preliminary diagramming stage.

Figure 7-17. Number, width, and length of the bars or lines connecting activities reflect the intensity of the relationship between activities.

The trials are usually done interactively on a computer screen or by using tracing paper, with refinements made on a new sheet. In the manual process, it can be very confusing when it becomes necessary to erase and modify existing diagrams. It is also extremely important to keep track of the number and type of relationships drawn. That is why it is better to do all the *A* activities first, count them, then do all the *Es*, count them, and so on. Some planners find it useful to color code the relationships to help keep track of them. (The conventions for color coding are shown in Figure 7-16.) It is also customary to use the standard ASME symbols used for process charting. Some planners use a turned-up arrow to represent offices, as the turned arrow looks like a house. Some planners use a turned-down *D* symbol or delay symbol to indicate service and support areas. Still other planners use only coded circle symbols for all areas, hence the term bubble diagram. The activity number is kept constant throughout the process—from activity area definitions, through relationship charting, through relationship diagramming, and so on.

Some areas (monuments) may be fixed beforehand and the planner should recognize this before beginning the diagramming process. Experienced planners frequently make use of a building plan or shell footprint and construct the relationship diagram directly on a copy of this shell.

An existing multi-aisle automatic storage and retrieval system (ASRS) or an expensive multi-machine flexible manufacturing system (FMS), for example, would be considered a fixed activity and its input and output locations would be used as activity area locations. This is a point that is frequently overlooked by new planners. The centroids of certain areas are not always the best areas for locating the relationship connection points for diagramming. For instance, in a long ASRS with input and output points at one end, the activity area symbol should not be placed at the center of the entire ASRS plan view. It should be placed at the input/output location. This type of situation often occurs when dealing with high-aspect-ratio or long departments where all materials enter and leave at one end. As will become clear later, when we add space requirements, it is not always best to work with the centroids of activity areas.

A word of caution, however: the ideal layout for a new facility that has no constraints should be done without preconceived thoughts as to the building shell. Theoretically, the best layout is

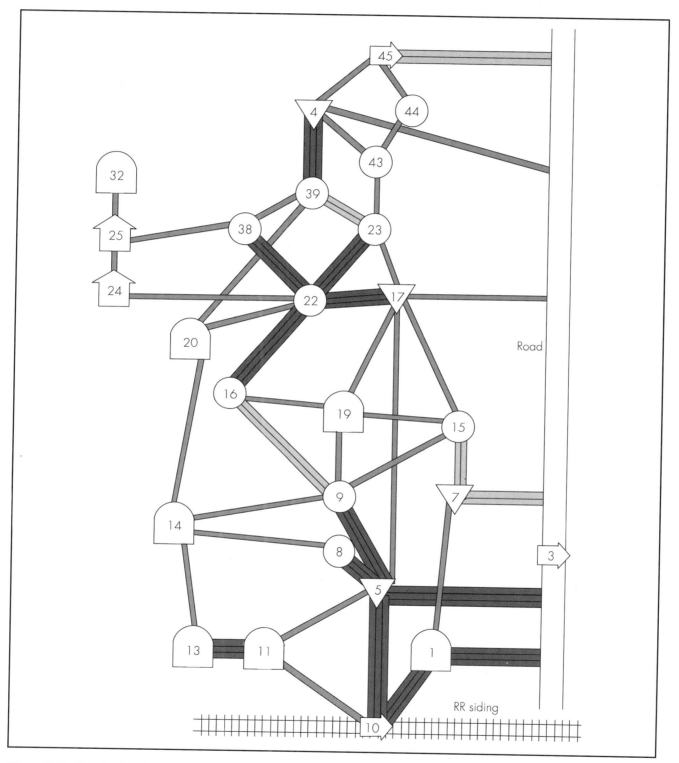

Figure 7-18. *Relationship diagrams somewhat resemble an airline's route map, with hubs (here, departments 5 and 22) representing heavily accessed hubs.*

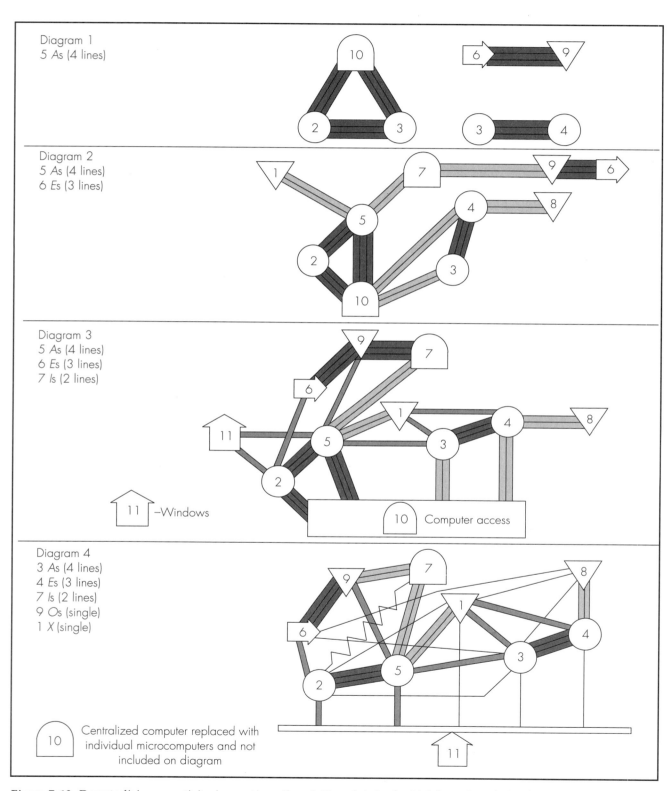

Figure 7-19. *Decentralizing an activity is sometimes the solution of choice for high-intensity relationships.*

achieved when there are no preconceived constraints. Such layouts or relationship diagrams can be modified at a later time if we are given a fixed shell or site to work with.

Diagramming relationships is usually the most difficult concept for new layout planners to comprehend. It takes time and practice to become proficient. Bear in mind that it has been mathematically proven that there is no one perfect relationship diagram. Likewise, there is no one perfect computer-generated solution for a plant layout. To construct such a diagram with 30 or more activity pairs of varying sizes would take many, many computers literally years, if not lifetimes to develop. However, with today's computer technology as an aid, we may be able to come close enough.

An analogy comes to mind about the difference between a scientist and an engineer. A bar of gold was placed at the end of a hallway 6 1/2 ft (2 m) away from a scientist and an engineer. Both of them could take 1/2 the distance of their prior step for each forward step they made towards the bar of gold. In other words, each was allowed to take one full pace or full step for their first movement. Their second step was 1/2 the distance of the first, and the third step was 1/2 the distance of their second, and so on. When asked how many steps or how long it would take to reach the bar of gold, the scientist said that "we are dealing with an infinite series and we will never reach the gold." The engineer replied, "I'll get close enough!"

We are only searching for a reasonable fit at this point in our analysis. The diagram itself will be adjusted later as space configurations and other constraints are taken into account.

Typical macro layouts for manufacturing plants have from 15 to 45 activities. Any more than that and the planner will surely be overwhelmed by the complexity of the problem. If you know for certain that two activities will be next to one another in the final layout, consolidate them into one area now to simplify constructing the relationship diagram. As previously discussed, there is no one perfect relationship diagram for a typical manufacturing plant, although there may be several good ones. The important fact to keep in mind is that the diagram selected will form the foundational

logic of the layout. If we have done a good job in establishing our activity relationships and we can hold those relationships when we later add our area requirements to the diagram, we should end up with several excellent and workable layouts.

TYPICAL COMPUTER-AIDED APPROACHES

Computerized approaches to layout planning can be grouped according to four major phases.

- Decision support systems.
- Computer-aided space planning.
- Computer-aided design (and drafting).
- Management information systems.

In this section, we discuss briefly the second category; the first category (DSS) was discussed earlier. The third and fourth categories are beyond the scope of this text.

As mentioned earlier, the most useful and comprehensive decision support tool for the layout planner is a computer-based spreadsheet and charting program. Another useful tool not mentioned previously is a CAD-based sketching/drawing program that shows areas of shapes (in real time) as the user moves, stretches, or elongates the shape. I have used inexpensive sketching programs with good results.

More sophisticated computer-aided space planning (CASP) packages are commercially available, but let the "buyer beware." Many are more useful in managing facility assets and in teaching students concepts rather than in real-world industrial applications. The sophisticated programs that can be used to actually develop useful, albeit crude, block layout alternatives are generally expensive. Normally, they are difficult to justify, unless you are planning a new facility or you are in the business of facilities planning.

Since the early 1960s, there have been literally hundreds of attempts by researchers and graduate students to develop algorithms for automatically developing plant layouts. Almost every respected university professor who has ever specialized in teaching facilities planning to students has tried to develop his or her own "best-fit" algorithm. Some programs are still used today in teaching environments and university settings. The better-known programs which thousands of industrial engineering students have either read about or used in their college careers are:

- CRAFT™ (Computerized Relative Allocation of Facilities),
- ALDEP™ (Automated Layout Design Program), and
- CORELAP™ (Computerized Relationship Layout Planning).

CRAFT is an improvement program (which means you need to input an existing layout into the program). CRAFT multiplies the flow, cost, and distances between the centroids of activity areas. It is actually the only program of the three that uses flow-of-materials data as the basis for proximity relationships. Some consistent flow rate must be used, such as pounds per hour, units per week, etc. The user is allowed to enter movement costs per unit distance. If this data is not universally available, a 1.0 value can be added to the matrix to make movement costs equal. Obviously this is not true in the real world, but at least an approximation is available. CRAFT generally has a limit on the number of activities involved (up to 40), but some of these can be fixed by the planner.

A total initial cost is computed, and then the computer examines changes in the layout. The program evaluates possible location exchanges between activity pairs which are either adjacent to each other or in the same area. This is repeated until no significant change is noted.

ALDEP is a construction routine, which means you do not need to start with an existing layout. The program randomly selects and places the first activity in the upper left corner of the layout sheet. The other relationships are then searched to find one (or the one) with the highest relationship score to the first activity. This activity is placed next to the first one and this process is continued until all the activities are accounted for. This process is done repeatedly for a number of layouts and the one that satisfies the most relationships is, theoretically, the best of the layouts generated. Since the first activity is randomly selected, many different layouts may be generated.

CORELAP is also a construction program. CORELAP uses some additional intelligence by calculating the activity which has the most rela-

tionships to other activities. In essence, CORELAP begins with a hub activity and places that area first in the center of the layout. An activity that must be close to the first one is then selected and placed next to it. The program searches for more *A* relationships and these are placed, then *E*s and these are placed, etc., until all activities are accounted for. CORELAP, unlike the other two programs, does take into account undesirable *X* relationships; however, they are treated in a purely theoretical, mathematical sense (which, as mentioned previously, can get the planner in trouble if he or she is not careful).

Sample representations of the layouts generated by these programs are given in Figures 7-20 through 7-22. The limitations of these programs

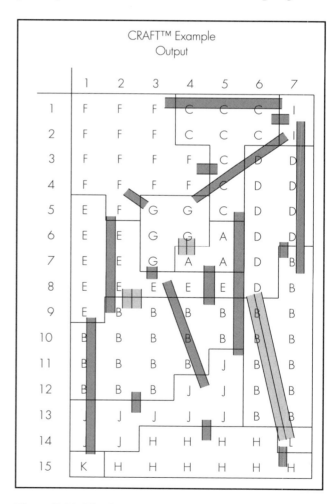

Figure 7-20. The CRAFT automated planning tool is an improvement program for existing layouts.

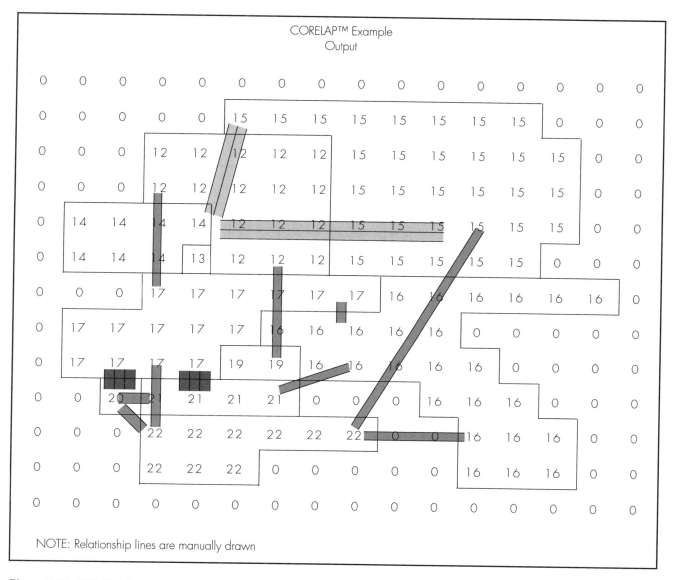

Figure 7-21. *CORELAP, a construction routine, needs no existing layout.*

are self-evident. Typical problems with automated programs include jagged wall constructions, the placement of high transaction areas like shipping or receiving in the center of a building, and other non-sensible results. However, sometimes they offer a starting point from which good layouts can subsequently be developed.

Unfortunately, the three programs mentioned are not very user friendly. None employ a true CAD-like graphics interface. The methods used do not handle realistic cost and distance factors

very well. Depending on which work center is placed first on some programs, different layouts will be generated, with each of them considered near optimal by the software.

Almost every university offering graduate programs in industrial and systems engineering have developed variations of the three most recognized older programs mentioned. There have been few commercial successes emanating from the universities. However, several good facilities planning programs have been

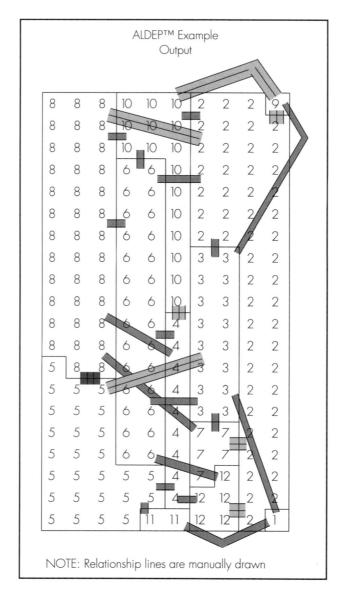

ALDEP™ Example
Output

8	8	8	10	10	10	2	2	2	9
8	8	8	10	10	10	2	2	2	2
8	8	8	10	10	10	2	2	2	2
8	8	8	6	6	10	2	2	2	2
8	8	8	6	6	10	2	2	2	2
8	8	8	6	6	10	2	2	2	2
8	8	8	6	6	10	2	2	2	2
8	8	8	6	6	10	3	3	2	2
8	8	8	6	6	10	3	3	2	2
8	8	8	6	6	10	3	3	2	2
8	8	8	6	6	4	3	3	2	2
8	8	8	6	6	4	3	3	2	2
5	8	8	6	6	4	3	3	2	2
5	5	5	6	6	4	3	3	2	2
5	5	5	6	6	4	3	3	2	2
5	5	5	6	6	4	7	7	2	2
5	5	5	6	6	4	7	7	2	2
5	5	5	5	5	4	7	12	2	2
5	5	5	5	5	4	12	12	2	2
5	5	5	5	11	11	12	12	2	1

NOTE: Relationship lines are manually drawn

Figure 7-22. ALDEP *starts with a hub at the center of the layout and works through the vowel/letter relationships radially.*

developed that can trace their roots back to graduate- level work of the universities.

Some of these new programs have resulted from modifications to the three programs mentioned. Typically, they have been developed by architectural firms, specialty engineering or consulting firms, and a few large CAD companies. All, however, are relatively expensive for the casual user. Unless a person generates crude block layouts for a living, these systems are seldom justified. Some of the programs do help with stacking of multistory buildings, but that, too, is a limited market, typically aimed at those architectural firms or institutions engaged in planning large offices. One of the better programs available for developing relationship (bubble) diagrams and stacking routines is WinSABA™. (See Figure 7-15 for an example bubble diagram developed using the WinSABA program.)

Some of the major CAD equipment suppliers provide a relationship analysis program with their architectural software packages. In the U.S., Computervision was one of the early leaders in this area. Autotrol and Intergraph Corporation also offered this type of analysis package at one time. These companies and their competitors should be contacted directly for what is currently available.

Appendix A of this book lists some of the more widely known and more useful personal computer (PC)-based software for developing relationship diagrams or macro block layouts. The list is not totally comprehensive, but does include those programs which were readily available in the U.S. at the time of this writing. Most of these programs require the user to input the activities to be included, the space requirements for each activity, and a relative closeness rating between activities. The majority of them use a routine similar to the graphical method we discussed in the previous sections. To learn more about these programs, contact the vendors; some offer demo packages.

BENEFITS AND PITFALLS OF USING RELATIONSHIP DIAGRAMS

One of the major benefits of using relationships to help develop plant layouts is their graphical and visual selling aspect. It is relatively easy to use the objectivity of the analysis and the color-coded, short graphic connections drawn between activity pairs to sell a layout to management or to obtain a buy-in from a reviewing group. However, the planner must continually keep in mind that relationships are only an approximation of reality. Remember that relationships are usually developed and ranked based on the *number of transactions or cube flowing* inside the plant (e.g., the number of fork truck trips between activity pairs). That measurement leads us to the major pitfall of relying solely on a relationship analysis approach in developing plant layouts.

Plant layouts based on relationships tend to focus heavily on materials transport cost and not necessarily true costs to the enterprise. Also, the number of transactions *may not be a totally accurate, relative measurement of actual materials handling costs*. Relationships between activities tend to lull one into assuming that all materials handling can be ranked on the same scale (e.g., an *A* relationship has the same *closeness* value, no matter what activity pair is connected). These relationships can be very misleading, particu-

larly if we change the mode of materials handling between activity pairs. In the final analysis, materials handling cost, cycle time, and the elimination of storage queues and nonvalue-added activities are the prime factors (not necessarily materials transport distance and transactions), in optimizing layouts. In our final layout plans, relationships are usually discarded in favor of an actual cost analysis and optimization based on the prime factors noted.

Pitfalls notwithstanding, benefits are usually achieved if the total number of nonvalue-added handlings, pickups, and setdowns are reduced. Those reductions, coupled with inventory reductions should be more important to the planner than simply reducing materials transport distances.

8

ESTABLISHING RELATIONSHIP DIAGRAMS FOR EXISTING PLANTS

As one would expect, there are major differences between establishing relationship diagrams for new plants when compared to already existing plants. Developing the diagrams for a totally new plant or site is usually an ideal situation. Conducting the investigations, data collection, materials flow analyses, and activity relationship analyses can be a truly rewarding and enlightening experience. This is particularly true for planners who have never done it before.

NEW PLANT VERSUS EXPANSION

It is my hope that sometime during your career, you have the opportunity to work on a project involving a totally new plant design. Whereas a new manufacturing facility will have (or should have) an ideal plan done once prior to the initial construction, usually several subsequent expansion plans will be developed for the facility throughout its lifetime. Therefore, on a *global* basis, there will always be more facilities planning projects involving existing plants.

In nearly all cases, developing relationship diagrams and conducting the entire plant layout process for an existing plant expansion or rearrangement is more difficult than for a totally new plant, the prime reason being the multitude of existing physical and attitudinal constraints the planner normally is forced to confront. Physical constraints are discussed in the next section; in this section, we highlight some of the other issues prevalent in developing changes for new plants.

PRE-DIAGRAMMING: PRACTICAL CONSIDERATIONS/ATTITUDES

Practical considerations generally focus on attitudinal issues, monetary limitations, timing constraints, and management structure and philosophy.

One of the major attitudinal and philosophical management problems facing many facility planners stems from a "separated" cadre of executive management. In this context, separated means an executive management team that is either located elsewhere within the corporation or, if on site, seldom ventures onto the shop floor. In these cases, when shop-floor visits *are* made, the visits are normally perfunctory formalities and no *real* knowledge of manufacturing processes is wanted or gained. Oddly enough, this is a management style that appears to be quite prevalent (and historically successful) in the U.K., U.S., and Germany, three of the world's most productive countries. The problem for facilities planners is that these far-flung executives are usually caught up in the buzzwords of the day with no real knowledge of the shop-floor impact of the buzzwords.

The drive to JIT manufacturing in the late 1980s and early 1990s is a prime example. There is no question that the JIT philosophy has been and remains a healthy goal for most manufacturers. The benefits of JIT have had an enormous impact on the balance sheets of corporations worldwide. However, the impact of JIT on *facilities* was never clearly understood by corporate executives. The drive to JIT left many shop-floor managers with ulcers and other stress-related maladies. True, with the huge reductions in inventory, everyone expected significant shop-floor space reductions, many of which were achieved. However, the *increase* in space requirements in *transactions*-related areas (receiving docks, incoming inspection, etc.) were almost totally overlooked by many corporations.

I remember working with one major automotive assembly plant supplier (with far-flung management) that finally elected to drop JIT deliveries (for its own company). Even though the company continued to ship to their assembly-plant customers on a JIT basis, the company returned *itself* to weekly and monthly deliveries on most parts. This was due principally to the backlog in their incoming inspection activities and the short time line for corrective action when vendor problems developed. If a supplier delivered daily, there was an inspection for every delivery "geared" into the company's computer-controlled materials system (which was managed from afar). Converting back to weekly and monthly deliveries cut the number of inspections, allowed ample time for corrective actions, and sped their process. Of course, larger safety stocks (inventory) were required with the weekly and monthly deliveries. The obvious longer-term solution appeared to be to return to JIT, use only certified suppliers, and to change the material control system software. Sounds simple enough. But after 2 years of effort, the company has only managed to certify one supplier. The company has still not been able to implement JIT on a large scale.

The effect of delivery schedules on space requirements and relationship diagramming is quite extensive. In a plant expansion, it is extremely difficult to plan for a JIT environment and then not be able to achieve it. You either have a storeroom or a WIP storage activity or you do not. Similarly, a plant or work cell either has an incoming inspection and hold/staging area or it does not. A relationship diagram either includes an activity or it does not.

Similarly, the drive towards "lean" manufacturing and the attempted elimination of buffer inventories on assembly lines and between work centers has caused extensive problems for many western manufacturers. As discussed earlier, even the founder of the "lean" manufacturing system, Toyota, makes good use of buffer inventories. Many western manufacturing firms believed for years that, if an individual operator on a Toyota assembly line could shut down the entire line, their operators should also have this capability (a dangerous assumption that ignored the beneficial effect of buffer inventories). In a similar vein, single-piece flow concepts have proven to be good in some situations and horrendous in others.

These portions of the lean manufacturing system were pushed by many western executive management teams with far more failures than successes. The usual result was a drop in productivity. Again, well-intentioned executive decisions from afar, without a base of manufacturing process and queuing knowledge, may lead to significant problems on the shop floor and significant problems with plant layouts.

In my experience, the successes of drastic manufacturing changes occur when the detail thinking and design is performed as close to the shop floor as possible. The overall direction or goal can and should continue to come from the executives above. As an example, it is far better for the executive team to push for "a manufacturing operation that reduces buffer inventories to the lowest (optimum) level consistent with"..., than to dictate "a manufacturing operation where all buffer inventories will be eliminated."

Unfortunately, as a facility planner, you may have very little influence over corporate directives or policies. However, corporate politics notwithstanding, you should definitely try to influence those directives that are obviously flawed.

Frequently, attitudinal problems emanate from lower levels as well as higher levels within the organization. Many people, particularly those operators who have held the same job or performed the same tasks for several years, may resent change, any kind of change that directly affects them. The best solution is to hear them out, take their advice where it is appropriate, keep them involved as part of the team, and train them in new techniques if at all possible. Training is the key point here. If the shop-floor operators can participate in developing the logic of the plan and they are trained in the methods of change, the probability of success is very high. Conversely, if they are not included in the processes, the probability of success is much lower, in many situations it will be zero. Some operators who feel they have been excluded from offering suggestions or ideas have a tendency to fight the proposed changes overtly and, in some cases, covertly. Usually the covert fighting of change does not manifest itself by tangible acts. Instead it is usually evidenced by a failure to inform the planners of critical process steps that have not been otherwise documented. Or failing to let the planner know about an obvious (to the operator) mistake or poor assumption the planner may be making. One of the fundamental data-gathering facets of any plant layout project is the interview process with operating personnel.

Other practical considerations include budgets for both money and time. We will all face monetary budget constraints and time budget constraints on facilities planning and plant layout projects. When planning new facilities, almost by definition there has to be a budget and task schedule in place. Unfortunately, this is not always the case when developing plant rearrangements and layouts for existing facilities. Many such projects languish for months and sometimes even years. This frequently happens when the responsibility is taken on internally as a part-time job by someone with more urgent daily responsibilities or when management does not have a clear grasp of the potential savings involved. Provided there are significant potential savings, it generally pays to engage an external professional consultant to develop the plans. The project will be completed much faster and be less costly than trying to do it internally part-time. An experienced "outsider" also may recommend improved cost-savings methods that the company has overlooked. Keep in mind, however, that internal resources are still required to work with the consultant during the data-gathering and analysis portion of the project.

Few companies can afford the internal staff and resources necessary to plan a major expansion project. Those that can are typically very large, high-growth organizations, capable of supporting such a specialized staff on a full-time basis with continuous facilities planning projects. It is normally counterproductive for a small or medium-sized company to attempt such a project using inexperienced internal personnel (with other major duty assignments) on an ad hoc basis. These part-time efforts generally result in cobbled-up poor layouts (placing equipment wherever there is an open space without regard to streamlining flow), high materials handling costs, and low morale. (An exception to this general rule emerges when the company must recruit new management to run the expanded portion of the operations. In that event, it is logical that the company have the new managers and other key shop-floor personnel required for the new expansion participate in the plant layout project.)

A prime *practical* consideration, which the planner needs to address as soon as possible, is the potential cost of relocating equipment and rearranging areas. The subject should be brought up for early discussion, particularly in a plant that is highly capital-intensive. Since the total scope of the possible changes is not known at this early stage, some of the cost questions are not easy to answer. A typical response from upper management usually is, "we will move any or all pieces of equipment as long as it is cost justified." However, in many situations it will be relatively easy to estimate the rearrangement costs but far more difficult to estimate the savings. Upper management is generally interested in "hard" savings which yield positive changes in cash flow. Motivational speeches and pep talks notwithstanding, upper management is seldom persuaded by *fuzzy* justifications such as *streamlined* materials flow, *improvements* in housekeeping, *increases* in employee morale, etc.

Even though you may be able to design a perfect, ideal, relayout of a particular area, the costs of moving large and expensive pieces of equipment and the downtime necessitated by the move may offset the savings. Moving many pieces of equipment can be relatively expensive if outside skilled trades such as millwrights and electricians need to be brought in for a lengthy plant and equipment rearrangement phase. This is one reason why it is so important to implement methods improvements in parallel with plant rearrangements. The methods improvements which yield "hard" savings can help offset the rearrangement costs.

With projects for existing facilities, one of the best things we can do is to try to bring budgets and schedules to the surface as early as possible in the planning process. Only when time and money are allocated against a formalized schedule can definite progress be measured.

EXISTING CONSTRAINTS AND MONUMENTS

Many plant layout projects of existing (in-use) facilities have physical constraints or built in monuments that must be dealt with. Some of the more common are:

- A need to save existing offices, lunchroom, restrooms, etc.
- A need to save existing truck docks.
- Low ceiling clear heights.
- Low floor-loading capacities.
- Cannot modify mechanical rooms (chillers, compressors, etc.).
- A requirement to use existing emissions stacks.
- Existing building column locations.
- Load-bearing walls and wind-brace locations.
- A requirement to isolate dirty operations against one wall.
- A need to salvage mezzanines or second-floor levels.
- Hazardous materials operations and storage.
- Fixed items: gold reclamation (cyanide) room, existing heavy processing area (e.g., painting, die casting, foundry, large punch-press, large drop hammer, coil slitting line, grit blasting), vibration-free area, cleaning pits, etc.
- Natural lighting.
- Direction of future expansion.

For the planner faced with these types of directives and constraints, it is best to fix those items or activities that cannot be moved in place on a rough sketch (see Figure 8-1) and develop or rotate the rest of the relationship diagram around the fixed monuments.

You should also recognize that, in some cases, monuments can be changed. For example, several land-locked factories that could not expand in any direction have had their roofs raised an additional 13 ft (4 m) or so *without disrupting ongoing production operations*. This allowed either a second manufacturing level (mezzanine) or a consolidation of storage into a high-bay racked area, freeing up ground-floor space elsewhere in the plants. The costs of the additional steel columns and wall structure and the erection costs to raise a roof generally ranges between 30 and 50% of new, single-level, pre-engineered building shell costs.

Similarly, a structural engineering analysis may show that selected building columns can be removed or relocated and replaced with an alternate supporting structure. Likewise, portions of load-bearing walls may be moved by incorporating a suitable alternative structure.

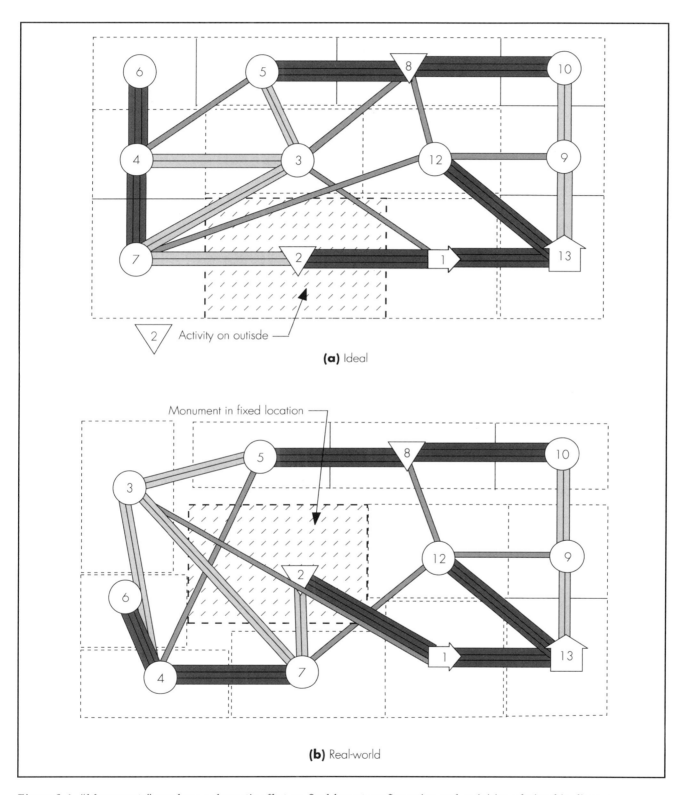

(a) Ideal

(b) Real-world

Figure 8-1. "Monuments" can have a dramatic effect on final layout configuration and activities relationship diagrams.

9

DEVELOPING THE SPATIAL RELATIONSHIP DIAGRAM

GENERAL CONSIDERATIONS

Based on our previous work in developing the relationship or bubble diagram and the projected space requirements, we are now ready to physically locate the activity areas.

To this point in our analysis, we were working only with relationships between activities. Now we are ready to take space configuration into account. Our goal is to retain our original relationships as best we can, while allowing for actual space requirements and special configurations.

On simple projects such as in departmental or work cell layouts, where noncomplex materials flow dominates, we can "hang" or place the space directly on the operation symbols shown on the operations process chart. This is the equivalent of the straight materials flow method we discussed earlier. On more complex projects, it is best to hang the space directly on the relationship diagram, as shown in Figure 9-1. It is also important to retain activity area identification and space requirements on each block of space.

Also critical at this point is to identify those activity areas that may have been neglected in establishing relationships. Typical neglected items or factors are miscellaneous underroof space, parking area locations, prevailing winds on truck dock locations, sunlight considerations, etc.

DIAGRAMMING FOR A "GREENFIELD" SITE

If possible, it is always better to work at least one alternative without constraints so that a truly significant layout improvement is not overlooked. A theoretically ideal (greenfield) layout should always be one of the alternatives developed for evaluation.

Experienced layout planners use cross-sectioned grid paper for establishing spatial relationship diagrams when working without the benefit of a computer, in which case, each square block represents a certain square area. However, a computer can be very helpful for this task. Several computer-based drafting/printing programs are available that allow you to easily move and edit defined area blocks for these diagrams. Some programs will even give you a real-time area readout as you change the shape of a particular area.

The general rules to follow in developing the diagram are:

1. Adhere to the same symbology and conventions used in previous steps.
2. Spread the activities apart to allow for large activity areas. That is to say, start with a magnified version of the original relationship diagram.
3. Make every attempt to "hang" the space requirements directly on the prior relationship diagram without changing any of the relationships to a great degree.

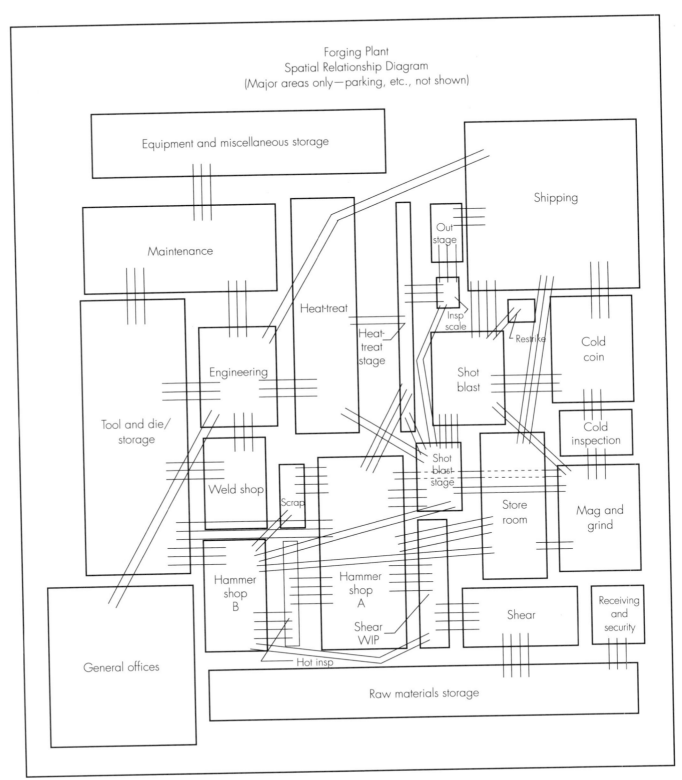

Figure 9-1. *The first step towards physically locating the activity areas is to "hang" the space directly on the relationship diagram.*

4. Take into account special or high-aspect-ratio configuration needs that are known at the time (e.g., a long piece of production equipment or machinery).

5. Avoid, if possible, relationship lines touching any other activity areas than the ones they are supposed to be connecting. If congestion occurs, draw the lines across unrelated areas in phantom or dotted-line fashion.

6. When the diagram is complete, color-code the hard copy for visual emphasis. It is best to stay with the standard color-coding procedure used throughout the analysis.

On complex projects, it is wise to make at least three alternative spatial relationship diagrams. These may be created using elements from other alternative relationship diagrams or may be derived from only one. The planner can use mirror images, different fixed locations for receiving and shipping, etc. Some changes will, of course, result in modifications to the original relationship diagram.

At this point, the planner should also be thinking about natural aisle patterns, future building column locations and other considerations, as well as the subject diagram.

It is important to keep in mind that, throughout the entire process thus far, the central objective has been to adhere to the logic pattern of our original relationships. These were based on our original flow and nonflow calculations. As long as we continually try to maintain both our space needs and their relationships to one another, we will end up with good, defensible plant layouts.

DIAGRAMMING FOR AN EXISTING PLANT OR PLANT EXPANSION

In some cases, fixed items or installations are in existence in a particular building. These are installations or monuments that management has determined *will not* be moved due to cost, sentiment, or any of a number of other considerations. The time to fix those in place is now, if you have not already done so in developing your relationship diagram.

The same general rules apply as in constructing diagrams for greenfield sites. Usually there are many more physical constraints in existing plants.

If we have honored all of the necessary relationships when developing the relationship diagram, we can draw the spatial relationship diagram separately, or in some cases directly onto the floor plan of the existing building shell plan or proposed facility floor plan. Just as with greenfield sites, on large projects with more than 20 activities, one or several cluster diagrams should be developed.

Of course, in developing the spatial relationship diagram and "molding" it into an existing building shell, the planner needs to be aware of physical building features such as columns, etc.

For small and large single-floor macro layouts (with 20 or fewer activity blocks), planners frequently skip the development of a spatial relationship diagram. For projects of that type, the relationship diagram can be overlaid directly onto the building shell (at a reasonable scale) and the alternative block layouts are developed directly.

Generally, it is recommended that a spatial relationship diagram be developed for multifloor projects. Even for multifloor projects with fewer than 20 activities, it helps to see the potential clusters of activities that may be placed on different levels. We discuss this more fully in a later section.

IMPORTANT PHYSICAL CONSIDERATIONS

Often, when trying to develop spatial relationship diagrams and macro layouts with existing plants, some important physical considerations may be overlooked or discounted. A prime example is varying ceiling clear heights within existing buildings.

If clear heights do vary, the planner must decide early on whether he or she will need to take advantage of the additional cube space available. The planner may have developed an ideal plant layout, only to discover that 30% more storage space is required because the ideal storage area falls within a low ceiling height area of the plant. The tradeoff or comparison evaluation will involve the value of the additional space required versus the operational materials handling savings in the ideal layout.

Similar problems may occur because of floor loading restrictions in certain areas of the plant or desired closeness to an outside wall. These types of problems are sometimes overlooked, or they may have inadvertently received a low affinity factor in the initial relationship development phase.

Future mezzanine areas also need to be taken into consideration. For example, perhaps a *boxed-in* area, that will need an expansion in a short time frame, should be placed in a high-ceiling-height area that can accommodate a mezzanine structure in the future.

Another example of where clear height is crucial is in work centers that need to be under an overhead crane. An ideal relationship diagram and spatial relationship diagram may show that several such activities are not located under, say, an existing bridge crane. To implement an ideal layout, the company may object to investing in a new crane and new steel supporting structures. Also, the planner must be sure not to place a noncrane-using activity, with the potential for a future mezzanine area, in or under a bridge crane area. This could be a particular problem if the noncrane-using activity separated two crane-using activities. Even though this appears obvious, I have seen at least three plants in which this mistake, intentional or otherwise, came back to "bite" the operation.

Another typical example would be where an ideal layout might show the relocation or shifting of all the machines in a particular department a short distance, say 10 or 20 ft (3 to 6 m). In those types of situations, the company may rationally elect to save the relocation costs and leave the equipment where it is. However, such a decision at this stage can have a disastrous "snowballing" effect on other dependent areas of the layout. The decision to leave equipment where it is generally results in a complete reconstruction of the relationship diagram and all further work with the subject department fixed as a monument, a reaction that may have far-reaching bad effects. It usually is the short-distance relocations that bring up these objection points. Oddly enough, if the analysis had shown an ideal layout with all of those machines moving a long distance, say 300 ft (91 m) away to the other side of the plant, there may not have

been an objection. Frequently, it is the *perception* that weighs more heavily than the facts. The perception by management in this particular case is that it is ridiculous to move equipment short distances. This is another reason why macro layouts should be backed up by a quantitative materials handling analysis. A computer-based quantitative analysis would allow actual operating cost comparisons to be made.

Similarly, some companies find it objectionable to move utilities such as water, sewer, and power lines—regardless of whether they are located above or below ground. Planners need to be aware of these types of idiosyncrasies before developing relationship diagrams or spatial relationship diagrams.

PLANNING NATURAL AISLE PATTERNS

The masters at planning natural aisle patterns are the Germans. Even small to medium-sized manufacturing plants in Germany tend to be extremely clean and well organized from a logistics point of view. In other western industrialized countries, such as the U.S., many small to medium-sized plants tend to be "kluges" of nonconnected activities with "weaving" aisles. In these situations, long, straight-running aisles tend to be the exception rather than the rule.

Obviously, straight-running aisles offer a much safer working environment than "weaving" aisles with 90° blind turns. As a rule, nonindustrial engineers generally do not recognize that there is also a cost penalty for "turns in the road."

For every turn in a materials handling aisle, there is an ongoing, perpetual added cost to the organization. Why? Consider the movement of a forklift truck as an example. Prior to reaching an aisle turn, the driver must decelerate to negotiate the turn. He or she will generally use the vehicle's brakes during this process. The driver will then slowly negotiate the turn and accelerate back up to normal speed before the same process occurs at the next turn. If there are two trucks traveling in opposite directions, one will usually come to a complete stop at a 90° turn to defer to another truck unless the aisles are extremely wide. The stopped driver will either wait for the other driver to negotiate the turn or, since both drivers have slowed or stopped, they may elect to hold a personal chat session at that point before starting up again. A similar situation occurs with people walking and moving pushcarts, pallet jacks, or handling devices.

Acceleration, deceleration, braking, and stopping all cost time and money. There is wasted driver time during these processes *and* the added wear on the trucks resulting in higher than normal maintenance costs. True, the costs may be slight, but accumulated over years, these costs may actually be significant and a considerable waste of resources. Consider a simple example of added labor costs alone. Suppose we have six forklift drivers operating in a medium-sized U.S. plant with nonstraight-running aisles. Let's assume their hourly pay with fringe benefits is $16.00. Let's also assume that they each average four round trips per hour and waste only 1 minute on each trip leg in negotiating nonstraight aisles (2 minutes total wasted time per trip). That is six drivers × 240 workdays per year × 8 hours per day × four trips per hour × $16.00 per hour × 2 minutes wasted each trip ÷ 60 minutes per hour = $24,576 wasted per year. Add the excessive truck maintenance costs, and the cost to the company is probably more than $28,000 per year. Is that our total costs? No, probably not. There also may be product or part damage costs caused by parts falling off vehicles negotiating the right-angle turns in the aisles. There also may be more than six drivers plying those aisles, wasting incremental amounts of time. Adding total costs in lost productivity in this example could easily bring our wasted time and costs numbers to in excess of $35,000 per year. If 5 years pass before the layout and aisles are straightened out, the company might just as well have thrown $175,000 out the window. Don't forget, we only assumed 2 minutes of wasted time per trip. I have been in dozens of plants where far more time than 2 minutes per trip was wasted due to nonstraight aisles. Even in the example discussed, almost an entire person's time (1,536 annual hours out of 1,920 work hours available) is wasted.

Obviously, the regulatory and liability impact associated with higher risks of accidents can make the lost time and added maintenance costs appear minuscule.

Of course all of this assumes that savings in the driver's time can be put to productive use. If straight aisles only serve to give the drivers an extra hour each day to read the newspaper, the company has far bigger problems than nonstraight aisles.

Sometimes, the space relationship diagram will help to give us an indication of heavy traffic aisles. Generally, where there are many relationship lines in a particular area, a major traffic aisle will be required. It takes some innovation on the part of the planner to develop natural and straight-running aisles. One shortcut is to sum all the areas of activities to be situated along the building shell wall. Divide this sum by the length of the wall to be used (minus cross-aisle widths); and this result will yield the constant width of each area along a straight-running wall. The length of each area can then be easily determined by dividing the constant width into the area required for each activity (see Figure 9-2).

Once a natural aisle pattern is established, the locations of fire doors and other access and egress points can be established to line up with the aisle pattern.

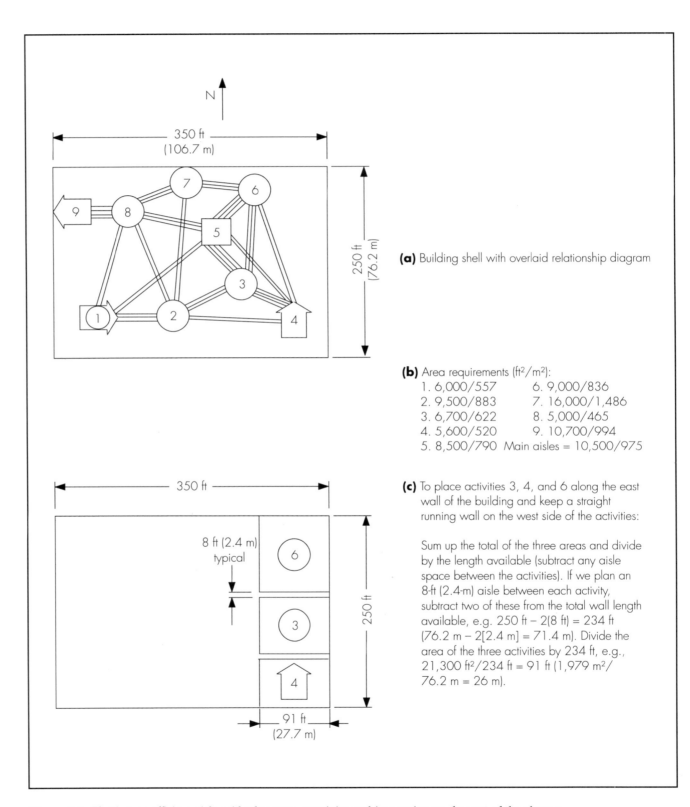

(a) Building shell with overlaid relationship diagram

(b) Area requirements (ft²/m²):
1. 6,000/557
2. 9,500/883
3. 6,700/622
4. 5,600/520
5. 8,500/790
6. 9,000/836
7. 16,000/1,486
8. 5,000/465
9. 10,700/994
Main aisles = 10,500/975

(c) To place activities 3, 4, and 6 along the east wall of the building and keep a straight running wall on the west side of the activities:

Sum up the total of the three areas and divide by the length available (subtract any aisle space between the activities). If we plan an 8-ft (2.4-m) aisle between each activity, subtract two of these from the total wall length available, e.g. 250 ft – 2(8 ft) = 234 ft (76.2 m – 2[2.4 m] = 71.4 m). Divide the area of the three activities by 234 ft, e.g., 21,300 ft²/234 ft = 91 ft (1,979 m²/ 76.2 m = 26 m).

Figure 9-2. Plotting an efficient aisle grid takes some creativity and innovation on the part of the planner.

10

DEVELOPING ALTERNATIVE LAYOUT CONFIGURATIONS

KEEPING YOUR EYES ON THE LOGIC AND THE ATTRIBUTES

When we have established our relationship diagram(s), area aspect ratios, and spatial relationship diagram(s), we can begin to "mold" the results into several workable layout alternatives.

Up to this point, we have discussed macro layouts of blocked areas of space. However, you should keep in mind that a similar relationship analysis procedure may be used for detail layouts. In detail layouts, instead of using blocked space, actual machine or equipment plan views are used for the blocks.

Thus far in the process, we have used a mechanically-oriented, logical, step-by-step approach. At this point, the experience and creativity of the planner should come into play.

Although the spatial relationship diagram may be a theoretically ideal layout in itself when closed together to form a block layout, what about the real world problems we run into in an existing building? For example, where are the building columns and the main aisles? Are main aisles as straight as they should be or are they joggled? (As we discussed earlier, joggled aisles cost money in materials handling labor and sometimes in property damage.)

And what about floor loading? Will the area we selected for steel storage handle the materials load as well as the fork truck loads? What about the plant's plumbing and drainage systems? What are our external transportation requirements—shipping and receiving docks, rail

sidings, etc. Instead of separating offices and welding by 100 ft (30 m), why can't we just erect a suitable wall between them?

As you can see, dozens of modifying factors can arise during the adjustment phase. Some of these may have been taken into account when we developed our spatial relationship diagram, others may have been neglected.

Also, major monument-type fixed installations such as automatic storage and retrieval systems, flexible manufacturing systems, and complex painting systems should not be placed in the center of a building or on a building face that is in the main expansion direction.

Although incoming utilities are not normally a major factor in block layouts, they most certainly can become constraints in some plants. This possibility should at least be checked by the planner. Maintenance departments are also sometimes neglected in layouts, but in some plants can prove to be extremely important locationwise. The handling of scrap materials can be another modifying factor.

Although these factors should have been taken into account earlier, if we have not done so, we need to do so now before finalizing block layout alternatives.

All of the foregoing material may be considered as modifying factors—step 10 on our pattern of procedures shown in Chapter 3, Figure 3-1. Typical factors that must be taken into account include:

- Existing building column locations,
- Facility clear heights,

- Existing load-bearing walls,
- Pits and foundation requirements,
- Hazardous operations (outside wall),
- Firewalls and underhung smoke screens,
- Fixed, in-place "monuments,"
- Natural aisle patterns,
- Access and egress routes and doors,
- Receiving and shipping docks/rail sidings,
- Floor conditions and flatness,
- Adequate separation or protection of "clean" areas,
- Utilities service entrances,
- Scrap handling and consolidation,
- Battery charging rooms/areas for electric vehicles,
- Location of mechanical room(s) (air compressors, boiler, chillers, etc.),
- Areas requiring ventilation,
- Areas that may require a mezzanine in the future,
- Shelving or horizontal carousel installations that may be double or triple stacked in the future,
- Tour routes and "image" for visitors.

If we already have established a building shell, we may need to fix the monument activities in place and rotate our spatial relationships around the monuments. If we do not have a shell available, and we are planning a totally new layout, we still need to be sure that common practices are adhered to. For example, computer programs that develop block layouts may have a tendency to place the shipping and receiving truck docks directly in the center of the building, with no direct access to an outside wall.

The planner needs to concentrate heavily on the materials flow logic and the original success attributes of the project. Working with the dozens of tasks and details involved in facilities and plant layout planning, it is easy to lose sight of the original attributes. The planner also must be wary of depending solely on relationship diagrams and spatial relationship diagrams for handling all of the layout needs. It is true that they are a tremendous benefit and they form a good selling tool. However, the ingenuity and experience of the planner is still required.

The two or three alternative block layout plans that have been developed to this point should differ enough among themselves to allow for an objective rating comparison. The method used to accomplish this is covered in a later section. All of the alternatives should allow for a review of relationships and how well they are honored. Some planners develop a new relationship diagram for each alternative at this point. These diagrams can then be compared with our original "ideal" to see how well we have satisfied our original inputs. Some of the computer programs available today will also score layout alternatives using weighted relationships and travel distance. In addition, software packages are available to help the planner calculate the materials handling costs associated with individual plant layouts. Many larger companies with adequate technical resources have set up their own spreadsheet-based computer programs for calculating materials handling costs.

The development of an example block layout alternative is shown in Figures 10-1 through 10-8.

EXPANSION FLEXIBILITY CONSIDERATIONS

One of the most important factors to bear in mind is the need for flexibility in the final block layout. The planner should avoid boxing in any activity that is sure to expand to a greater degree than its surrounding areas. These areas should be placed near an outside wall on a building face that lends itself to further expansion wherever possible.

To establish those activity areas that are most likely to expand in the near future, an *expansion sensitivity analysis* should be performed. Ideally, this analysis should be completed *before* developing relationship diagram(s). If completed at that time, you can use relationships as a tool to help plan for flexibility.

To accomplish the expansion sensitivity analysis, you must set a time frame target for the *next* expansion (not the current expansion being planned). This is usually established by upper management and is based on growth projections. A typical target would be 5 years after the current expansion, unless growth needs dictate a need for additional space prior to 5 years in the future. In Chapter 6 we covered ratio trend analyses and growth projections for space. Using those projections along with the operating process charts as guides, you need to review the

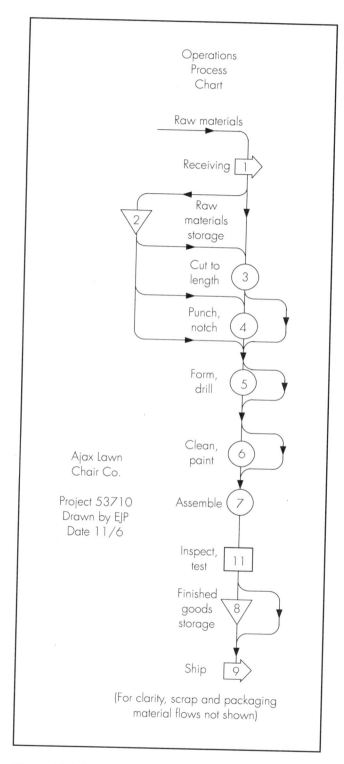

Operations
Process
Chart

Raw materials

Receiving 1

Raw
materials
storage 2

Cut to
length 3

Punch,
notch 4

Form,
drill 5

Clean,
paint 6

Assemble 7

Inspect,
test 11

Finished
goods
storage 8

Ship 9

Ajax Lawn
Chair Co.

Project 53710
Drawn by EJP
Date 11/6

(For clarity, scrap and packaging
material flows not shown)

Figure 10-1. Development of a block layout begins with an operations process chart that maps the major elements in the manufacturing flow.

various manufacturing processes to determine where the bottleneck conditions or "choke" points are located. These are the areas that will absolutely need more space as production demand increases. You also need to consider how the choke points in the process may change. These changes may occur as new products or equipment come on stream or as local areas are modified before the next physical "brick-and-mortar" expansion. Needless to emphasize, this analysis of changing choke areas is not an easy task. It will take a considerable amount of investigation to compile a logical listing of probable expansion areas and how those probabilities may appear just prior to the next (5 years out?) expansion. However, if you want to consider expansion flexibility, as you definitely should, these probabilities must be developed.

Similar to the ranked bar chart for the "intensity of materials flow," a bar chart should be developed for the *probability of expansion* for all underroof activities (see Figure 10-9). Usually this is done from a macro layout or block layout perspective. Those choke activities that absolutely require additional space with production increases are ranked at the top of the chart. Those with a low probability of requiring additional space fall at the bottom of the expansion probability chart. If a highly detailed analysis is performed, you can expect a typical Pareto-type of expansion probability distribution. If gross estimates are used in lieu of a detailed analysis, a more stepped type of chart would be developed.

The planner can now rank these probabilities with the vowel/letter ranking system—*A*, *E*, and *I* as shown in Figure 10-9. In this case, what you are actually defining is the relationship of the listed activities to an expandable building wall on the shell of the building. Using these "expansion flexibility" relationships in the relationship diagramming efforts should prevent the bottleneck areas from being located in a boxed-in, central portion of the facility. These relationship "pulls" or "ties" are somewhat different than the ones discussed in prior sections in that these relationships may be drawn to *several* walls (if the plant has more than one direction of potential expansion).

It should be noted that we do not want to give the impression that this is an overly simple

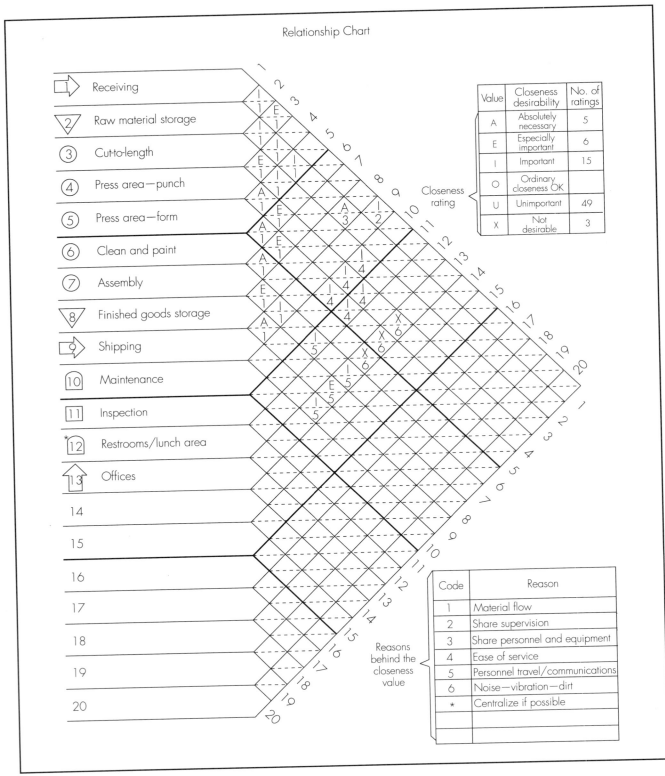

Figure 10-2. *The relationship chart identifies the activity "closeness" priorities and the reasons for the priorities.*

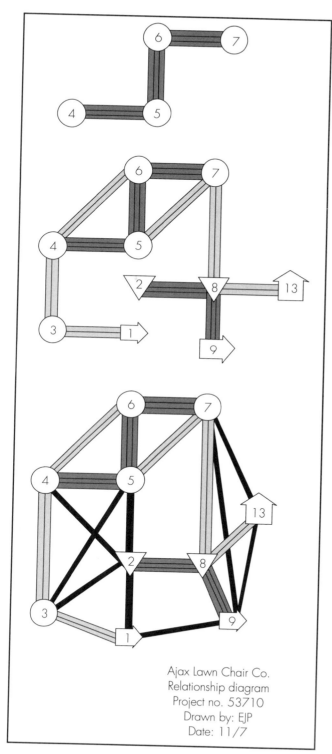

Figure 10-3. *Relationship diagrams are developed using graphical rules for the lengths of the relationships, and are run through several iterations by the layout planner.*

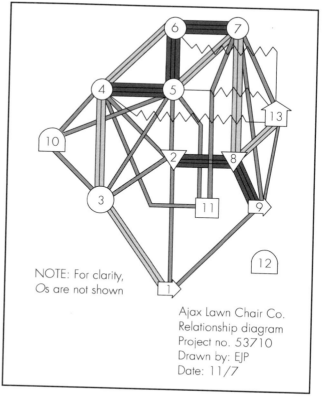

NOTE: For clarity, Os are not shown

Ajax Lawn Chair Co.
Relationship diagram
Project no. 53710
Drawn by: EJP
Date: 11/7

Figure 10-4. *After several "what-if" iterations, a working version of the relationship diagram emerges.*

analysis. For instance, let's presume the relationship diagram correctly shows the highest probable bottleneck (activity *A*) to be placed next to an expansion-facing outside wall. Let's also assume that the second highest bottleneck operation (activity *Y*) has absolutely no relationship to the first bottleneck (e.g., two different product lines). During the next expansion, the highest probability activity may be moved entirely into an expansion space added to the outside of the wall. That would leave a hole for the expansion of the next most probable bottleneck. Therefore, the second most probable bottleneck (activity *Y*) may need a *strong expansion relationship to the first activity* and not necessarily to an outside wall. Instead, activity *Y* may need to be close to the space occupied by the first bottleneck. Since this can be a highly complex procedure, you should conduct this analysis concurrent with the development of a comprehensive master facility plan.

Activities Listing and Area Requirements		
Department Name	Ft², m² Required	Aspect Ratio
1. Receiving	15,000/1,394	
2. Raw materials storage	20,000/1,858	
3. Cut-to-length line	40,000/3,716	4 to 1
4. Press area } punch and notch	30,000/2,787	
5. Press area } form and drill	62,000/5,760	
6. Clean and paint	40,000/3,716	4 to 1
7. Assembly	76,000/7,061	
8. Finished goods storage	40,000/3,716	
9. Shipping	25,000/2,323	
10. Maintenance	13,500/1,254	
11. Inspection	4,500/418	
12. Restrooms/lunch area	4,500/418	
13. Offices	21,000/1,951	
Total	391,500/36,372	

NOTE: Allowances for main aisles are included in ft²/m² required.

Ajax Lawn Chair Co.

Figure 10-5. *When a logical relationship diagram has been developed, a listing must also be compiled of all major activities in the plant and the space needed for those activities.*

In most situations, a more simplified approach may be used instead of vowel/letter rankings. I use a special "Ex" relationship for those one or two activities that will positively need an expansion during the next expansion cycle. The "Ex" relationship is established between the bottleneck activity and an outside wall that faces the direction of expansion.

With new plant or new expansion designs, you may be able to avoid many of the normal constraints associated with building-column locations. It is highly recommended that clear-span construction be considered wherever possible. Given suitable climatic and seismic conditions, pre-engineered metal buildings with clear spans of up to 115 ft (35 m) or so generally add less than 2% to construction costs. The layout and rearrangement flexibility offered by clear-span construction is generally well worth the added cost.

INTRADEPARTMENT/CELL DETAIL LAYOUT CONSIDERATIONS

Detail layout planning follows the same general pattern of procedures as block layout planning.

However, the level of detail information required is much more equipment—and operator—specific.

Detail planning is normally performed for a given, fixed-space configuration that has already been determined in the block layout phase. Again, it needs to be emphasized that the planning process should progress from the site plan to the block or macro layout plan and then to the detail plan. There will always be overlap between these phases. Also, modifications will usually, but not always, be required within the macro and detail layout phases to ensure an overall acceptable and workable plan.

The detail layout phase of the layout process requires considerably more time and therefore generally costs more than the block or macro layout phase. Although detail planning requires a relatively high level of mechanical aptitude and skills, it is sometimes delegated to members of the company who are at a lower level in the organization with little training in industrial engineering or "motion economy." This is probably because a closer interface must exist between top management and the individuals planning the

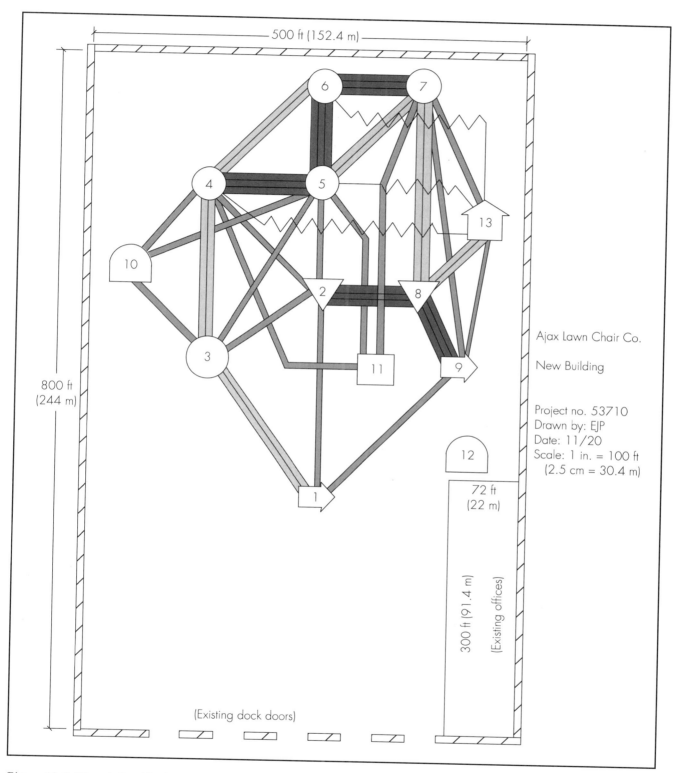

Ajax Lawn Chair Co.

New Building

Project no. 53710
Drawn by: EJP
Date: 11/20
Scale: 1 in. = 100 ft
(2.5 cm = 30.4 m)

Figure 10-6. *The relationship diagram imposed on the footprint of the facility shell provides the layout planner with a workable map of the production flow and relative intensities of flow among several activities.*

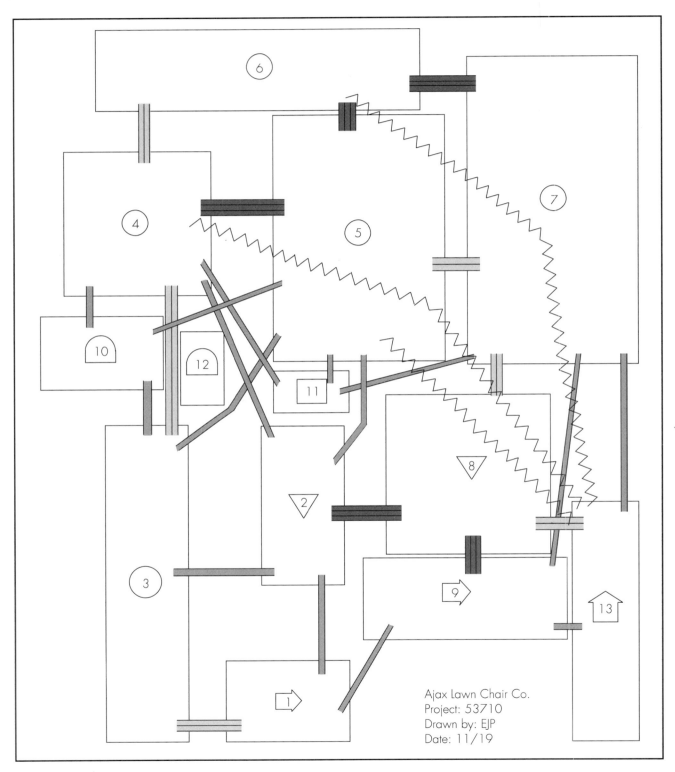

Figure 10-7. *Further refinement of the layout produces a spatial relationship diagram in which the "blocks" become more clearly defined.*

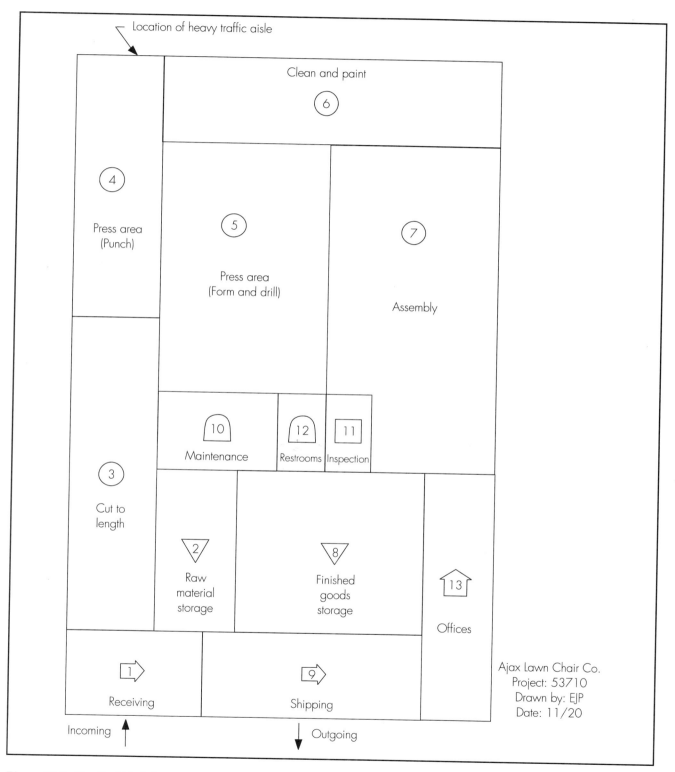

Figure 10-8. *The final block layout shows, for the first time, the plant-floor sizes and shapes molded into a logical layout.*

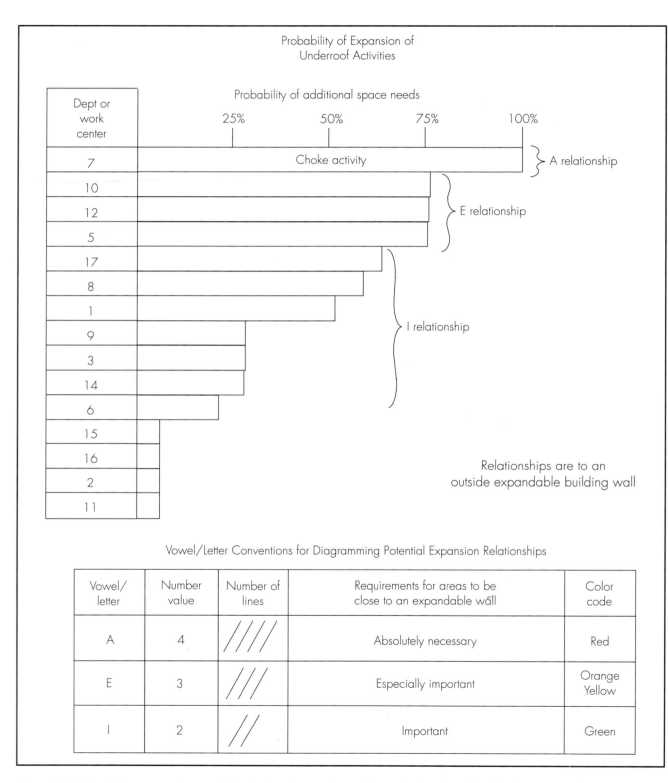

Figure 10-9. In a future expansion analysis, the bar chart and vowel/letter designations identify the urgency of expansion space needs by department or work center.

site and block layouts. This interface is normally not required at the detail level unless a very sophisticated system is being installed and upper management wants to keep close to its progress.

A heavy emphasis on coordination is required between individual area supervision, the shop-floor team, and the detail layout planner. As mentioned, in many cases the shop-floor team develops its own layout. It is preferable that an experienced manufacturing engineer or industrial engineer be a part of the layout team. If that is not possible, some members of the team should definitely have had an adequate level of training in motion economy, ergonomics or human factors, and industrial engineering techniques.

The mechanical process or logic used in detail layout planning within specific departments or cells is quite similar to the process we have been describing. Instead of relationships being developed among blocked-out areas, they are now developed between individual operations and machines. In the detail layout, the location of specific pieces of equipment such as workbenches or cabinets of supplies is usually included in developing relationships. Incoming utility service locations, runs, and drops are also very important in this phase.

Operational process charts or routing sheets are almost a necessity for developing intradepartmental or intracell materials flows. Each worker's space, equipment, and materials setdown areas have to be accounted for. It is helpful at this stage of the process to have "library parts," showing the plan-view equipment outlines and power needs, pre-programmed in your CAD system for individual pieces of equipment. If a manual drafting system is used in lieu of CAD, equipment templates become quite useful.

Instead of using activity symbols for establishing the relationship diagram or blocked areas for the spatial relationship diagram, you now use the individual equipment templates or library parts themselves.

Adjustments to the block layout areas may be required during the detail layout phase. These normally develop due to improvements noted in the detail layout if a slight modification to the overall layout is made. As previously discussed, the block layout phase and the detail layout phases should always overlap.

Fundamental questions you should ask in developing detail layouts are:

- How will materials be delivered to and taken away from the detail area, and where will the pickup/setdown points be placed? Can delivery and take-away points be combined to get "double duty" from materials handling trips?
- Are all materials within easy reach of the operators? Have we taken human factors into account? Have we designed our station to take advantage of the force of gravity?
- Will specialized lifting or rotating equipment be required to pick up or move materials within the workstation?
- Have we considered *all* of the physical requirements (as well as costs) associated with accessories, such as full (and empty) materials containers, pallets or totes, tooling, toolboxes, chairs, benches, ergonomic aids, piping and ductwork, lighting, cabinets, racks and shelving, scrap or discrepant materials containers, carts, battery chargers, personal belongings, etc.?
- Will we ever need a mezzanine level or materials drop level above the workstation? Do we need to allow for this potential in the future?
- How will power, water, compressed air, communication lines, or other utilities be delivered to the detail area and where will the drops or below-floor service points be placed? *Will there be flexibility for changes?*
- If returnable containers are to be used, where will they be placed and how will they be returned? If containers are not used, how will trash and scrap be handled and removed from the area?
- Are there any wasted walking steps, reaches, or other motions? If so, are the detrimental effects minimized?
- Has the layout itself caused excessive operator idle time? Can the layout be revised or turned to make good use of any slack time?
- Can an operator perform one or more additional tasks instead of only those which have become customary?
- Can all equipment be serviced and maintained properly? Do we have to physically move pieces of equipment to pull shafts or other items from another piece of equipment? Will

the servicing of one piece of equipment adversely affect another operation within the workstation? Will access or materials handling vehicle paths be blocked while servicing any piece of equipment?

- How many nonvalue-added material handling pickups, travels, and setdowns are occurring? Can the number of nonvalue handlings be reduced by changing the layout?
- Are we absolutely sure that no queue or additional storage space is needed within the workstation? Or if needed, *have we allowed enough room and flexibility for change or adjustment?*
- Is there any unnecessary worker lifting, reaching, pulling, pushing, pinching, wrist-turning, or bending? If so, the ergonomics of the layout need to be addressed and modified.
- Are any operators "trapped" inside a detail work center layout without a means of quick and safe egress? If so, the layout needs to be addressed and modified.
- Did an objective "devil's advocate" critique the layout before it was cast in stone? Question the sequence and timing of operations within the work center. Can we combine these operations with another operation or work center and increase total productivity or space utilization? Are we performing the operation "because we have always done it that way"? Is there a better way to do it?
- Have we simulated the actual conditions closely enough to be assured there are no safety problems? Have we designed in the proper safety shutdown points, machine guards, etc.?

As with the block layout phase, it is very important to keep all supervisors and other interested parties informed of the detail layout process. Significant changes should receive signoffs from managers and supervisors.

Approvals for detail layouts should be received from:

- The individual supervisors responsible for each workstation within the area,
- Safety and/or security personnel,
- Supporting services departments,
- The person in overall charge of the operations covered by the layout.

Some companies also insist that the shop-floor operators who must work within the layout have the opportunity to express their feelings and offer improvement suggestions. Although the operators normally do not have signoff responsibility, they should be able to provide their inputs to the layout planner. Ideally, this should be accomplished in the detail layout concept stage and before completion of the detail layout.

The acceptance signoff procedure should come as a result of company policy and not an individual's desires. It is important that the procedure be followed so that it does not appear that the planner is trying to put an approver on the spot by signing-off the drawings.

ERGONOMICS/HUMAN FACTORS

Human factors engineering and ergonomics play a key role in the design of detail layouts and workstations. Human factors engineering, per se, focuses on proper workplace design for maximizing productivity. In a partnership role with human factors engineering, ergonomics is more heavily focused on avoiding work-related injuries. Ergonomics, in a manufacturing context, can generally be defined as *the study of the interaction between humans and their tools, equipment, production materials, and workplace environment.*

Table 10-1 shows typical anthropometric data used in the U.S.

If operators are in a standing position (e.g., picking materials from shelves), there is a "sweet" zone for maximum efficiency. The sweet zone for picking materials from shelves is the point where all (standing) pick-and-place activity is *between the nose and the knees.* In comparison to the sweet zone:

- The average pick rate above the worker's nose is two times slower.
- The average pick rate below the knees is three times slower.

The sweet zone notwithstanding, selecting and picking items at the extremes of the sweet zone (e.g., at the nose and knee level) may not be sound ergonomics. Bending to select and pick materials, particularly at the knee level, is not recommended. Likewise, reaching, selecting, picking, or any work effort that forces the worker

Table 10-1. Placement of Controls (Anthropometric Data)

Location	Men	Women
Maximum reach (measured from shoulder pivot)	24 in. (61 cm)	21 in. (53 cm)
Normal reach (measured from elbow pivot)	14 in. (36 cm)	12 in. (30 cm)
Best standing control zone (between shoulder and waist height)	43 to 57 in. (109 to 145 cm)	40 to 53 in. (102 to 135 cm)
Best sitting control zone (between shoulder and waist height)	26 to 40 in. (66 to 102 cm)	24 to 37 in. (61 cm to 94 cm)
Assumed standard work heights (to accommodate 80% of population)	17 in. (43 cm)	16 in. (41 cm)

NOTE: These figures are based on population studies in North America and may vary for other regions of the world.

to raise his or her arms above the shoulder level on a repetitive basis is not recommended.

Since people vary in physical size and stature, it is highly recommended that height-adjustable equipment be employed wherever possible. For example, worker's chairs should be capable of rotating as well as having a height adjustment feature. Some of the basic empirical "rules" of ergonomics follow:

- Set the height of a workstation (where the operator's hands will be working) approximately 2 in. (51 mm) below the elbow position. This usually requires adjustable-height workstations to suit multiple worker heights.
- Use a flexible, rubber-like padded foot surface for operators who must work in a standing position. This allows better blood circulation in the lower leg and foot areas.
- Develop and enforce a common-sense safe lifting and bending policy which includes posture recommendations. In the U.S. the National Institute for Occupational Safety and Health (NIOSH, located in Cincinnati, Ohio) publishes a work practices guide for manual lifting.
- Given two poor manual materials handling alternatives, pushing is preferable to pulling. More workplace injuries are associated with pulling than pushing heavy objects.

- Use a swiveling work positioner or easily-rotatable table for manipulating workbench parts or products.
- Avoid repetitive reaches. If repetitive reaches must be made, they should be no more than 20 in. (51 cm) from the body, preferably less.
- Avoid elevating the elbow above the shoulder. Also avoid moving the elbow away from the front of the body. Keep reaches as short as possible.
- For seated operators, avoid any reaches below the seat or waist level.
- Avoid standing reaches behind the body centerline where the torso must be twisted. These types of reaches are acceptable for seated operators *only* if they are seated on a chair that swivels and the reach does not require twisting of the torso.
- Avoid workstation designs that require flexing or twisting of the wrist. Never design a workstation that requires repetitive flexing of the wrist more than 15 degrees (and preferably less than 10 degrees) away from the palm.
- Avoid any repetitive wrist bending or rotation that requires movement of greater than 5 degrees towards the thumb or more than 15 degrees towards the little finger.
- Avoid repetitively rotating the hand or forearm more than 45 degrees from the elbow pivot point.

- Avoid subjecting operators to
  - Repetitive "pounding" with the palm,
  - Repetitive "pinch" types of finger gripping,
  - Improperly-sized handle gripping (handle gripping should be as muscle stress-free as possible).
- Avoid a workstation design that places an operator in an environment with excessive and repetitive vibration.

It should be noted that these empirical rules are not based on any scientific studies or governmental standards or directives. The rules are based strictly on my experience and opinion and are offered as guidelines only. Governmental regulations will obviously take precedence. The reader is urged to check the latest governmental regulations to ensure compliance.

Figures 10-10 and 10-11 show a few of the ergonomic aids available to you that will help to prevent cumulative trauma or repetitive motion types of injuries.

Ergonomics and workplace design have become extremely important in recent years. Although the focus on improvements in workplace design has been gradually evolving since World War II, the primary push towards a much heavier emphasis started in the mid- to late 1980s. During that time, statistics on the costs of workplace injuries were much more publicized than in prior periods.

The cost factors associated with an injury to an experienced worker can be enormous. Typical costs include:

- The loss of the employee's skills, dedication, and experience during the healing process;
- The pain and suffering the injured employee must endure;
- Hospital and home care costs;
- Convalescent care costs;
- The additional wages that must be paid to a less skilled individual (assuming the injured worker continues to be paid);
- Potential legal representation and court costs, particularly if long-term disabilities are incurred;
- Increased insurance premium or worker's compensation costs;

- Training costs for the new replacement employee;
- The costs associated with the lower level of production attributable to employing a less skilled worker;
- The administrative time and costs associated with government regulations and injury tracking and reporting. These may be coupled with potential monetary fines and excessive physical remedies dictated by governmental regulatory agencies.

In the mid- to late 1980s, the U.S. witnessed an explosion in reported work-related injuries. The most notorious of these was *carpal tunnel syndrome*, involving inflammation of the tendons in the hand affecting thousands of workers. Many other repetitive stress disorders or "cumulative trauma" injuries also began to surface. These injuries are called "cumulative" because they do not surface overnight. They are the result of very small (and frequently unnoticed) injuries, typically accumulated over a long time period. Eventually the injuries become acute, with sometimes chronic and incurable results.

The latest information shows that, despite a decline in the total number of workplace injuries, repetitive motion injuries in the U.S. increased approximately 7.5% between 1992 and 1993. The annual reported cases in the U.S. climbed from 281,000 to 302,000 during that period. Actually, these types of injuries are only a small fraction of total workplace injuries. In 1992 alone, some 13 million workers claimed back injuries caused by ergonomically-poor work environments.*

It is far more beneficial for employers to aggressively insist on a properly engineered workstation than to accept even minor injuries as unavoidable. For many years, too many companies considered the costs of minor injuries just one of the unavoidable costs of doing business. However, the tangible and intangible costs (morale, image, reputation, etc.) associated with workplace injuries can far exceed the 10 items listed here. A safe working environment pays for itself many times over.

---

*According to the publication: *Business First*, CBJ Publications, Inc., Columbus, Ohio, November 6, 1995.

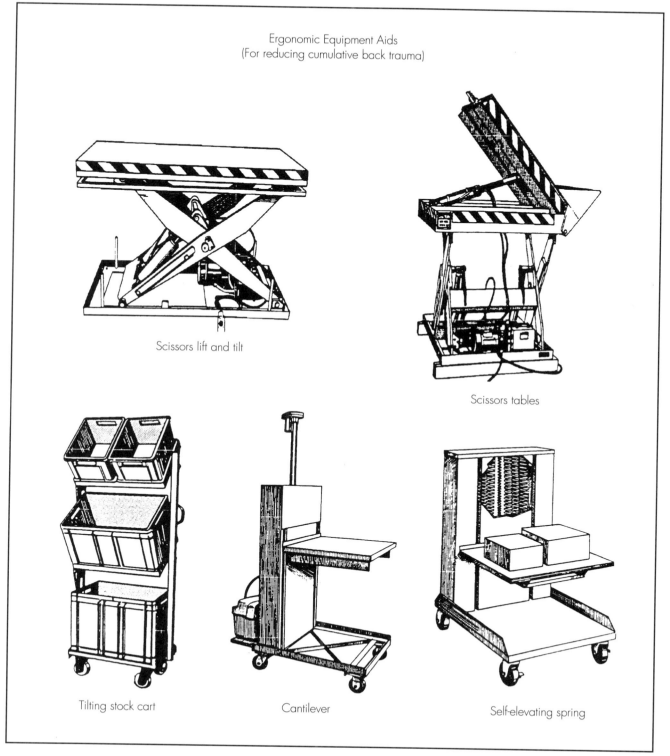

Figure 10-10. *Crucial to the development of detail layouts are human factors considerations that are not only mandated by governmental agencies, but serve as basis for sound and sensible design. (Courtesy Vestil Manufacturing Co.)*

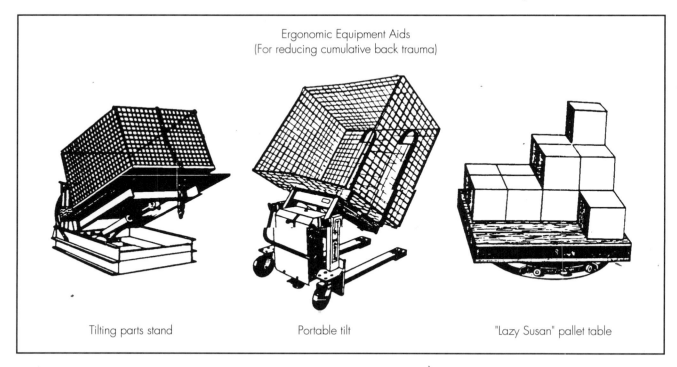

Ergonomic Equipment Aids
(For reducing cumulative back trauma)

Tilting parts stand   Portable tilt   "Lazy Susan" pallet table

Figure 10-11. Entire industries have emerged dedicated to the design of ergonomically friendly manufacturing aids, producing devices such as these. (Courtesy Vestil Manufacturing Co.)

BUDGET LIMITATIONS

When developing new or rearranged plant layouts, one must always be cognizant of budget constraints. For our purposes, the *budget* actually has three aspects, one dealing with cost limitations and the other two dealing with schedule and human resource limitations. Schedule and human resource limitations are unique to each project and are beyond the scope of this book. However, some very helpful PC-based software tools are available that can help you develop and manage resource schedules and costs. The most widely used and comprehensive project management software packages are listed in Chapter 3.

A considerable amount of judgment must be exercised regarding plant layout project costs. When working on an existing plant layout rearrangement, you need to be wary of becoming overly idealistic. *It is usually much easier to develop an ideal layout than it is to cost-justify the factory rearrangement.* This is not so much a problem with new plants, but it is definitely a problem with existing plants. And it is particularly true for those plants that are highly capital- and equipment-intensive. A typical example is shown in Figures 10-12 through 10-16.

This metalworking plant produced forged products in several different buildings on a relatively large site. Special equipment with special foundations were included in each building. Many, but not all, of the existing pieces of equipment were, surprisingly, *not* considered "monuments" and could be moved. The company required an economic justification (and rightly so) for any proposed rearrangement. After an extensive analysis of the company's operations, an improved layout was developed which would significantly reduce materials handling effort with a minimum impact on production equipment movement. The new layout also would allow a reduction in the number of fork-lift trucks (those involved in materials handling) from 30 to 17. The before-and-after material flow patterns are shown in Figures 10-15 and 10-16, respectively.

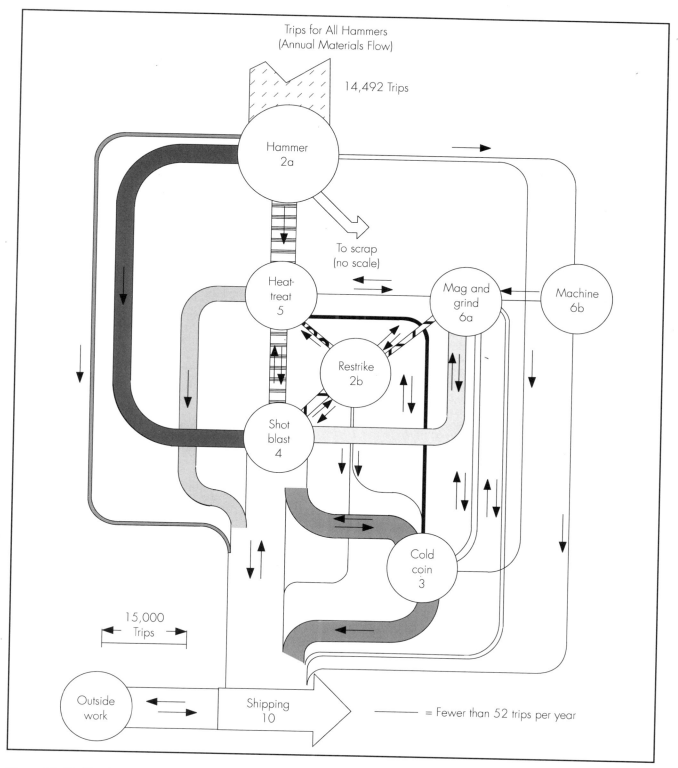

Figure 10-12. *Plotting the intensity of current materials flow (forklift truck trips in this case) is a key first step in justifying a proposed plant rearrangement.*

Materials Handling/Relationship Report By Product			
Product name: ALL **Production volume: 1/Year**			
From-to	ft/m	$	Relationships
Between LG HAMMER SHOP and HOT INSPECTION			A 4
Between SM HAMMER SHOP and HOT INSPECTION			A 4
Between RAW MATERIAL STORAGE and SHEAR			A 4
Between LG HAMMER SHOP and SCRAP		145,274	A 4
Between WIP and LG HAMMER SHOP		96,770	A 4
Between SHOT BLAST STAGE and SHOT BLAST	11,737,102/3,577,469	78,606	A 4
Between SHOT BLAST and SHIPPING	13,026,808/3,970,571	60,196	A 4
Between SHEAR and WIP	3,662,000/1,116,178	51,912	A 4
Between COLD COIN and SHIPPING	12,201,028/3,718,873	51,214	E 3
Between LG HAMMER SHOP and HEAT STAGE		46,124	E 3
Between LG HAMMER SHOP and SHOT BLAST STAGE		44,104	E 3
Between SCALE and OUTSIDE STORAGE (OS) STAGE	7,800,418/2,377,567	38,058	E 3
Between OS STAGE and OS INSPECT	6,905,660/2,104,845	34,852	E 3
Between WELD SHOP and TOOL AND DIE STORAGE			E 3
Between 9000 and MAINTENANCE			E 3
Between HAMMER SHOP and MAINTENANCE			E 3
Between HAMMER SHOP and TOOL AND DIE STORAGE			E 3
Between MAINTENANCE and TOOL AND DIE STORAGE			E 3
Between STORE ROOM and SHIPPING			E 3
Between WELD SHOP and MAINTENANCE			I 2
Between LG HAMMER SHOP and STORE ROOM			I 2
Between SM HAMMER SHOP and STORE ROOM			I 2
Between COLD COIN and MAINTENANCE			I 2
Between SHOT BLAST and MAINTENANCE			I 2
Between HEAT-TREAT and MAINTENANCE			I 2
Between MAINTENANCE and STORE ROOM			I 2
Between OS STAGE and OS HEAT-TREAT	4,811,904/1,466,668	27,354	I 2
Between OS HEAT-TREAT and OS STAGE	4,676,276/1,425,329	26,868	I 2
Between SHOT BLAST and COLD COIN	5,408,934/1,648,643	26,838	I 2
Between SM HAMMER SHOP and SCRAP		26,686	I 2
Between HEAT-TREAT and SHOT BLAST STAGE	4,718,286/1,438,134	25,196	I 2
Between WIP and SM HAMMER SHOP		24,564	I 2

Figure 10-13. *By enlisting the aid of a computer and software designed specifically for materials handling analysis, the planner can calculate materials handling costs. A sample output page is shown here.*

A computer-based before-and-after materials handling cost analysis, using FactoryFLOW™ software, was performed (see Figure 10-13 for one page of the output). The results of the analysis showed an annual total materials handling cost savings of just over $310,000. Operating with the new layout would yield the same num-ber of "handling" transactions but would reduce travel distance significantly. Components of the materials handling savings included reduced materials handling labor (driver) costs; reduced fuel, power, and maintenance costs on the truck fleet; and reduced lease and depreciation costs for the truck fleet. However, the estimated costs

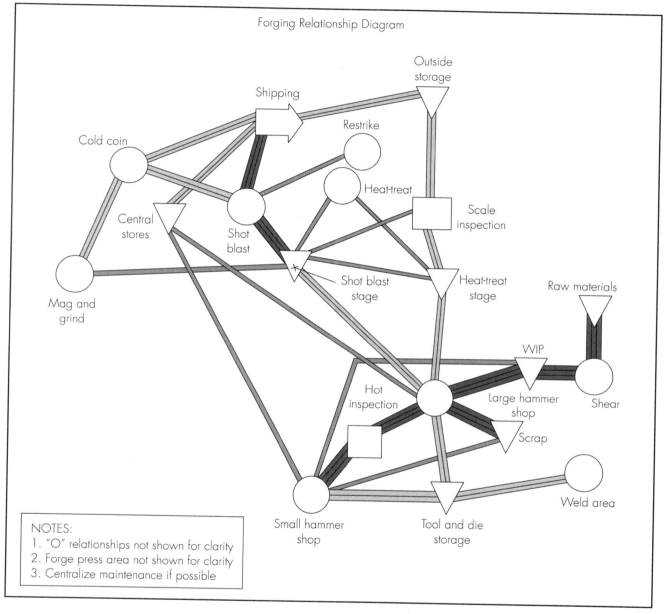

Figure 10-14. *The relationship diagram establishes the "best-case" scenario for the proposed layout. From this plot, the planner can begin to design the optimum physical layout for his or her plant.*

of building renovations and moving the equipment necessary to achieve these savings exceeded $1.7 million. The company could not afford to make all of the short-term changes necessary to achieve the savings. Instead the company used the new layout to form a long-term master site plan. This allowed a gradual longer-term planned move towards the benefits that could be achieved with improved layout. The results of the study were extremely useful in determining where a new, very large piece of equipment should be located at the site. The complex, state-of-the-art equipment system had been on order for more than a year and was due

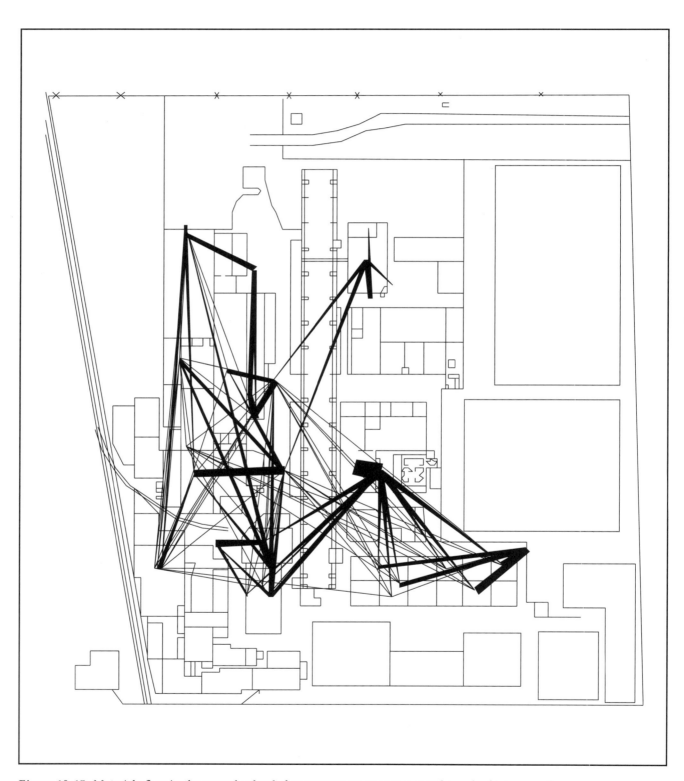

Figure 10-15. *Materials flow in the example plan before rearrangement was a complex web of intense pathways.*

Figure 10-16. *The proposed – or "after" – materials flow network shows a simplified and more efficient flow of materials moving far less distance at less cost.*

to arrive within 2 months. This new equipment, once set in place, would truly become a "monument" that would never be moved in the future. The study proved to be an invaluable aid in determining where this new equipment would be placed.

As this example demonstrates, one of the points to keep in mind is that the total costs of factory rearrangement do not always have a short-term payback. Generally, labor-intensive operations which do not have an extensive amount of sensitive, very-high-cost, capital equipment installed can be rearranged at a relatively low cost. Rearrangements of existing capital-intensive operations are more difficult to justify. Obviously, if there is a very high wage scale within the plant, and the plant has an extremely poor layout to start with, changes are more easily justified. The importance of getting the layout right *at the beginning* cannot be overemphasized.

Let's assume our example company was operating at a 9.8% pretax U.S. profit level. That is to say, every incremental $1 million in sales revenues generated $98,000 in pretax profit. Had the metalworking plant done nothing, and a new competitive plant was built with the same processes but using a better layout, the old plant would be at a severe competitive disadvantage. The new competitor would have at least a $310,000 annual pretax advantage. At the same ratio of expenses to sales, the old company would have to generate an *additional* $3.16 million in sales revenues just to *stay even* with the new company in pretax profits. And this is just one result of a poor layout. There are other hidden costs as well: high product and part damage, low employee morale and productivity, missed shipment schedules, lost materials about the plant, potentially higher inventories, and so on.

Also, when considering rearrangements to an existing plant, the planner should not try to focus solely on reducing travel distance between activities. Certainly that is a major component of costs but transactional and inventory costs may be even more important. In the example just discussed, we did not change the process or the number of handlings, per se; we only changed the distance or "transport costs" between activities. As discussed earlier, relying totally on relationship analysis, which in itself relies almost totally on transport costs, may not offer the best solution.

When budgeting the cost of a rearrangement, you should strive to improve both processing and materials handling methods as an integral part of the total justification package.

Some of the more traditional cost items associated with existing plant rearrangements include:

- Engineering and phased-plan development costs. These also may include the cost of required architectural work. Move sequence/method/ schedule planning should be analyzed.
- New construction or existing building renovation costs—lighting, plumbing, electrical, mechanical, painting, floor coating, etc.
- Potential cost of lost production during the rearrangement (may require "build-ahead" inventories). Rearrangement may also incur the cost of temporary, off-site storage facilities (usually leased).
- Millwright and other skilled trades costs for equipment disassembly, movement, reassembly, and utilities connections.
- Relocated equipment downtime costs—you may need to make some use of outside production resources for a short period of time (during the equipment move and debug time period).
- Granting of a building or renovation permit may require that some portions of an older building be improved to current building and fire code requirements. It may require building improvements to allow for handicapped or disabled workers (whether the company currently employs disabled people or not).
- Affected environmental emissions permits may need to be reissued.
- Internal coordinator/project manager resource costs.

MANUFACTURING CELLS

CELL HISTORY

Prior to the 1930s, the "industrial revolution" in the U.S. and Europe was focused primarily on increasing production of consumer products. Most manufacturing companies were small in size (and still are today), but lacked extensive capital and equipment resources. They did, however, have an ample supply of relatively low-cost labor.

Manufacturing organizations of that era were in awe of the Ford example of mass production. As mid-sized companies were growing, the inherent problems of mass production became evident to a few, particularly those producing many engineered or machined parts. The scheduling problems and high WIP inventories were obvious to a few enlightened companies. However, since most companies were experiencing similar situations (i.e., it was a way of life), the negative effects of mass production in large batches were not well publicized.

Experiments with breaking up large, traditional batch-oriented shops into more efficient flow-through cells began in Europe in the 1930s. Agricultural equipment manufacturers originally experimented with cells for producing a wide variety of machined parts. As an example, a typical cell would manufacture most of the shafts produced for several tractor models. It was evident that higher labor productivity and focus could be achieved with a cellular approach, usually at the expense of lower equipment utilization. But cells were still oddities to most companies. Since higher cell capital costs were perceived as a problem, the cell philosophy did not gain a great deal of interest prior to World War II.

The basic cell characteristics are:

- They are relatively small (however some may be defined as entire plants).
- They consist of two or more operating workstations independent of each other process-wise, but integrated or coupled into a series of sequential operations.
- The cell produces a completed testable object. It converts materials into a product or family of products. Group technology (GT) cells may produce similar components and not complete products, per se.
- Workstations generally have some capability to change fixtures or tooling to handle a family of products.
- Normally, cells are scheduled as a complete business unit. They operate within a "pull" manufacturing system.

Interest in cells further waned during World War II as military equipment manufacturing, on a massive high-volume scale, was implemented (see Figure 11-1).

Immediately after World War II, mass-production techniques in the West were still in vogue. These techniques were used to mass-manufacture consumer products in high volumes to fill the demands of war-rationed populations. The tremendous baby boom after the war sparked a very high rise in family formation. All of the consumer needs that are inherent to starting and growing families drove most of industry towards mass-production techniques using large batch sizes.

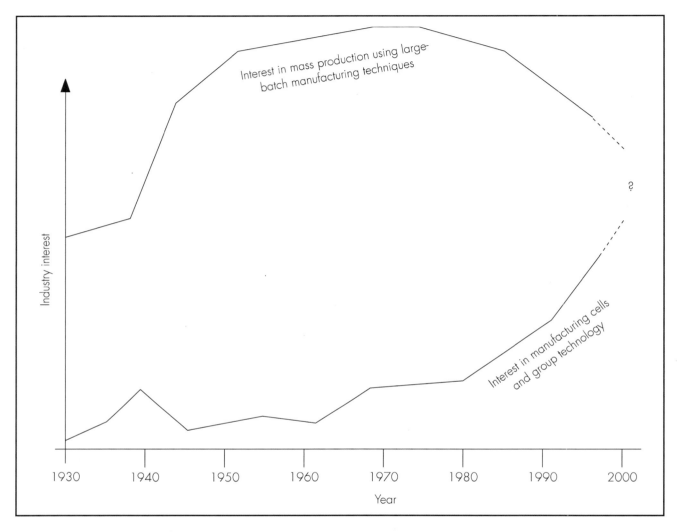

Figure 11-1. *Interest in manufacturing cells following World War II was relatively slow-growing until the 1980s, when the scope of manufacturing became global and the demands of customers became more exacting and unpredictable.*

Immediately following the war, however, some limited interest was beginning to again develop in cells, both in Europe and in the United States. They were not called "cells" at the time, but they were cells by today's definition. Early examples of cellular layouts appeared in U.S. publications in the late 1940s and early 1950s.

As Japan started to rebuild its industries after the war, its industrial base lagged far behind the United States. However, Japanese industry was very conscious of waste, much more than the industrialized Western countries. Japanese manufacturers were quick to recognize the materials savings that could be gained from cellular production techniques. Since Japan was receiving immense amounts of new capital infusions from the West, there was heightened interest in cellular production techniques.

As Figure 11-1 depicts, there was renewed interest again in cellular production in the mid-1960s through the early 1980s. This interest coincided with the growth of numerical control (NC) and computer numerical control (CNC) machine tools. This advancement in technology led a rush by large Western manufacturers to implement multimillion-dollar investments in flexible manufacturing systems (FMS). Many of these systems were funded by the Western countries through government contracts as part of the cold-war defense buildup.

Although FMS is still used by many large defense contractors, the interest in these huge (expensive) systems decreased considerably after the mid-1980s. Coinciding with the decreased level of interest was, and still is, a large buildup of interest in cells (but for a different reason) which somewhat negated the effect of decline in expensive FMS.

During the 1990s, interest in cells in the West has been spurred by two dynamics: increased demand for "custom," small-lot deliveries of product directly to consumers, and time-based competition.

As an example, in prior years, large mass-producers would manufacture and ship truckloads of product on pallets to regional distribution centers, say, once a month. Those centers, in turn, would ship pallets of products to the various outlets of merchandisers, generally twice a month. Now the trend is for the outlet to order a pallet of mixed products, packed in a particular sequence, directly from the factory *and only when needed*, with 48-hour delivery (effectively bypassing the middleman distribution center).

Likewise, in some countries in Asia, (India, Japan, etc.) there was no huge base of regional distributors as existed in the West. Industries in these countries were often forced to build only what was needed, when it was needed. They could not afford, nor did they have access to, distribution channels and warehouses that are taken for granted in Western countries.

Customer demands for more choice in product selection, better quality, and speedier deliveries have become increasingly important factors to most manufacturing industries. Recent industry trends indicate:

- A change from scheduled "push" approaches to demand "pull" approaches to manufacturing.
- A philosophy of "variety" manufacturing with quick changeovers, building only enough parts/products to meet current demand and no more.
- A gradual change from large-lot inventoried parts and products to small-lot, JIT production.

- A trend towards *single-piece flow* and *cellular* production techniques.
- Large portions of factories rearranged by product or families of products rather than being laid out by process.
- Much greater emphasis on team-building, quality circles, and employee empowerment and accountability.
- Much greater emphasis on setup and inventory reductions.
- More manufacturing organizations being structured to focus on particular products or business units rather than on traditional functions or processes.
- More emphasis on core competence and "focused" factories.

Coupling the factors of changing demands of customers with the need to better manage (and reduce) inventory assets in the Western industrialized countries has sparked renewed interest in manufacturing cells and group technology.

GROUP TECHNOLOGY

Rising and falling interest in cellular manufacturing has mirrored the same type of interest pattern in group technology (GT). This is partially due to the use of GT in designing efficient cells for large manufacturing organizations. Typically, large companies using GT produce a wide variety of complex but similar parts.

For example, a manufacturer of a hundred different models of agricultural tractors might manufacture hundreds of different cylindrical shafts of many different lengths and diameters within the plant. Some of the shafts might require milled keyways, necked and threaded ends, precision bores, or internal and external splines, etc. In a traditional plant layout by process, the company might have a lathe or turning department, a milling machine department, a screw machine department, a boring mill department, a gear shaping department, and so on. These departments would traditionally be separated by function and expertise into different areas of the plant. This type of layout generally forces the company to use a batch-oriented scheduling approach. This approach, in turn, results in large "tubs" of parts (WIP inventories) being transported and stored throughout the plant. As discussed earlier, this type of manufacturing approach also can result in high costs and poor quality. In an attempt to optimize throughput, lower shop cycle times, lower costs and improve quality, a different type of cell may be implemented.

The cell is focused on producing similar component parts or "families" of parts from different products. In the case of the agricultural equipment manufacturer, a shaft manufacturing cell would be developed. The cell would be designed to include a completely mixed set of production equipment, from lathes to gear shapers. The layout of the cell would be optimized by process sequence so that much smaller batch quantities could be run through the cell with very little materials handling and very small WIP inventories. In essence, the parts continually flow, almost in a straight line by process sequence, through the cell from start to finish without ever leaving the cell. Completed shafts of many different sizes and with many different machined features emerge from the cell. Obviously, the typical down side of this type of scenario is the relatively low equipment utilization levels of the machines used within the cell. Since some machines may be used on only a small number of parts that the cell handles, these machines would have very low utilization, and thus a higher capital investment is normally required. Also, although the scheduling appears to be much easier, setup costs and setup times are usually increased due to the shorter lot or batch runs of any particular part.

To optimize these conditions, many large companies implement computer-based GT techniques to select the parts to be included in a cell. Basically, these techniques use a rigorous part-coding scheme, with each part assigned a unique code number. Along with a digit for the part material, the subsequent digits represent particular part geometry and physical feature requirements. For example, by sorting on the part codes in a computer data base, you, as a planner could very quickly select all cold-rolled steel shafts between 1 and 3 in. (2.54 and 7.6 cm) in diameter, between 10 and 20 in. (25.4 and 51 cm) long, with a threaded end and milled keyway. Further sorts would allow you to optimize the particular attributes and machining needs of a

particular grouping of parts that allow an optimum cell layout to be developed. Although conceptually sound, the major problems with GT coding schemes are the extensive amount of time and costs involved in initially coding all of the parts in a multitude of bills of materials. Typically, only large companies (usually those with deep pockets or government contract backing) can afford to spend the time and force the discipline to keep the coding system going. Frequently, part and product changes, as well as subcontracting needs, tend to make layouts based on group technology somewhat less flexible than may be required in today's manufacturing environment. GT does, however, play a very important role in large design engineering organizations.

Another benefit of coding schemes accrues to engineering and purchasing departments. Coded parts schemes are extremely useful to product design engineers in helping them to select common parts (that have already been designed) rather than reinventing the wheel each time they design a new product. For instance, in the agricultural example given, before the product design engineer designs a new shaft for a new model tractor, he or she could very quickly search the coded data base to determine if an existing part could be used instead of an entirely new part. This use of GT (and manufacturing cells) can have positive internal impacts in several areas because it:

- Frees up engineering resources to do other things;
- Reduces unnecessary part proliferation (reduces paperwork and file system needs);
- Lowers overall inventories and reduces storage space needs;
- Reduces new part procedures (vendor certifications, extensive testing, etc.);
- Simplifies materials purchasing, control, and scheduling procedures;
- May lower overall materials costs;
- Increases equipment utilization;
- Simplifies computer-based warehouse management systems needs;
- May help to simplify and streamline shop-floor layouts.

GT affects more people and procedures within an organization than manufacturing cells. Like cellular manufacturing, GT is an "old" technology. However, its use will probably become more prevalent as personal computer use becomes even more common in industry.

GT coding system interest peaked in the 1950s, then peaked again in the 10-year period between 1976 and 1985. Figure 11-2 shows an adapted coding system originally used in the mid-1960s which is still valid today.

A GT coding scheme for cylindrical parts is shown. If a large company could GT code all of its thousands of machined parts, it would be relatively easy to design manufacturing cells. The cells would be based on classes of materials and machining similarities. Figure 11-3 shows a typical GT coding system for sheet metal parts.

Likewise, GT coding schemes also allow product engineers to easily select existing parts with the right dimensions, moment of inertia, etc., rather than design entirely new parts for every new product model. GT coding technology offers many more benefits which are beyond the scope of this book. Suffice it to say that GT can be highly beneficial in designing manufacturing cells in companies that produce many complex but similar product models and parts. On the negative side, the expense and time associated with analyzing and GT coding an existing base of parts can be very high. High implementation and upkeep costs have been the prime GT coding detractors, which have prevented widespread use of the technology.

GT also has been used as an aid in helping to select the proper mix or "grouping" of machines or equipment to produce similar parts. There have been several attempts to computerize this process, but most grouping analyses have been done on an iterative basis using tabular processes. The basic thrust of these types of analyses is to determine the most efficient groupings of pieces of equipment. However, there are far too many variables to make this a simple process. Labor variances, operating cycle differences, setup variances, space limitations, people issues—and, most importantly, flexibility—make this too complex an issue for solving by a purely tabular approach. Innovativeness of the layout planner and the cell operating team is still the prime factor in achieving cell layout success.

THE THREE MAJOR *P* FACTORS IN CELL PLANNING

The following three factors affect every plant layout in general and every cell layout in particular:

- Process issues,
- Procedure issues,
- People issues.

As depicted in Figure 11-4, people-related issues far outweigh the first two categories. The level of importance of people issues in cell design cannot be overstressed. People issues include:

- Policy and labor-management issues,
- Involvement in cell planning,
- Cultural change,
- Attitudes,
- Egos,
- Emotional considerations,
- Cross-training needs,
- Instilling multidisciplinary teamwork,
- Environmental issues,
- Communication needs,
- Cell organization and leadership,
- Buy-in.

A cell may operate effectively with less than perfect procedures and processes, but it cannot operate effectively, if at all, without *all* of the people issues being resolved. Resolving people issues *must* be accomplished during cell design and *before* implementation. In a traditional large-batch, process department-oriented layout, people-related issues are very important, but one person with an attitude problem will not shut down production. In a cell, where each operator is *immediately* dependent on his crew members, each operator becomes a key person and everyone is dependent on one another.

Procedural issues include:

- Operating procedures changes,
- Scheduling and scheduling control,
- Quality assurance needs,
- Materials handling,
- Materials management,
- Purchasing procedures,
- Maintenance,
- Unit (cell) accounting,
- Performance reporting,
- Communications procedures.

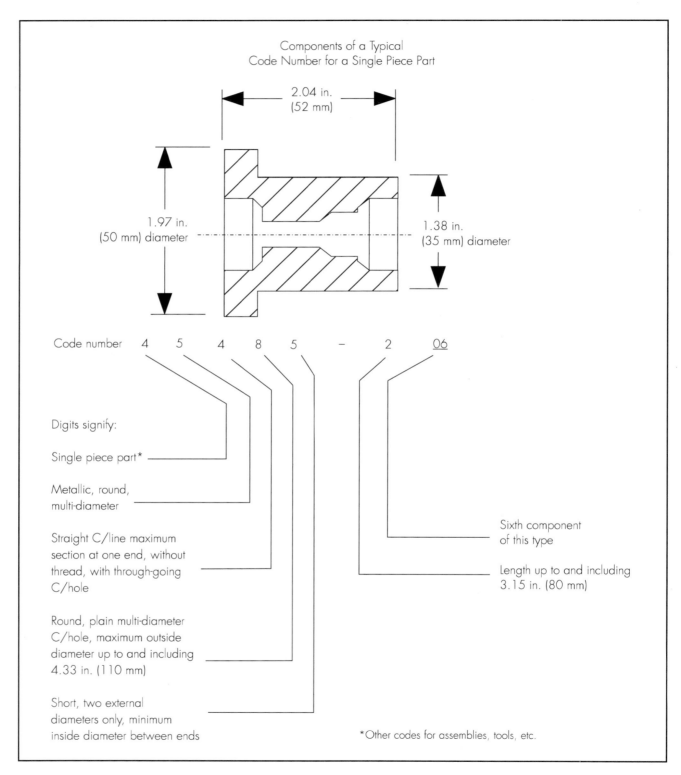

Figure 11-2. *Part coding schemes, like this one from the mid-1960s, identify all the key attributes of a single part, facilitating design, production, and purchasing functions.*

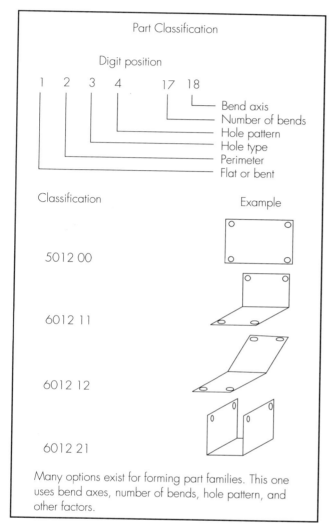

Figure 11-3. By classifying parts based on materials and machining similarities, and GT coding them, companies could easily design manufacturing cells to produce them efficiently. The cost of implementing coding schemes, however, can be prohibitive. (Courtesy Forming & Fabricating *magazine.)*

Process issues include:

- Facts, data, projections;
- Engineering;
- Physical controls;
- Analysis;
- Equipment;
- Tooling;
- Physical location and layout;
- Utilities distribution;
- Flexibility for change.

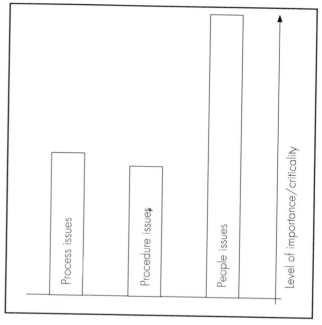

Figure 11-4. Factors impacting all layout and cell designs involve processes, procedures, and people. People issues — outweighing the other factors by a wide margin — run the gamut from governmental regulations to personal idiosyncrasies like ego, attitude, and emotion.

For an effective cell layout, all of these factors have to be addressed. All issues are interrelated to some degree, particularly communications and reporting. All issues from each of the categories must be solved together. One team must be involved in the entire planning, not three separate teams working on three separate categories.

REDUCING NONVALUE-ADDED OPERATIONS

One of the main targets of any cell or plant layout plan should be the reduction of the number of nonvalue-added operations. Normally, achieving that goal is one of the prime benefits of a manufacturing cell. Figure 11-5 shows just a small portion of a typical part's life within a factory. You will note that in the dozen or so steps shown, only one adds actual value to the part.

Even the time to perform the step that states "perform operation" is not total value-added time. Figure 11-6 repeats the steps in Figure 11-5 and depicts how much time can be classified as

Tracing a Typical Part's Life
Through the Factory

1. Receive raw material   — ——

2. Move to incoming inspection   — ——

3. Move to storage/staging   — ——

4. Place in storage   — ——

5. Remove from storage   — ——

6. Move to first operation (stack)   — ——

7. Pick up part   — ——

8. *Perform operation*

    (only value-added operation)   — ——

9. Place in outgoing stack   — ——

10. Move to WIP storage   — ——

11. Repeat steps 4 through 10   — ——

12. And so on   — ——

Figure 11-5. Value-added assessment of all production activities is the logical first step in eliminating operations that add cost to the product but no value.

nonvalue-added in the process. Breaking down the very small part of time that makes up the step(s) "perform operation" shows that only a small portion of that time is actually adding value to the part. One can conclude that, in traditional large-batch-oriented process department types of layouts, the great majority of time is wasted in queues, handling, and storage. A well-designed manufacturing cell reduces to a minimum wasted time, wasted handling, and bloated inventory storage.

CELL PLANNING PROCEDURE

A step-by-step planning procedure for manufacturing cells is shown in Figure 11-7. It begins with executive management communicating the vision and championing the process.

At the very beginning of the cell planning process, executive management should set up a steering committee, as shown in Figure 11-8. The two groups should then define:

- The purpose of the cell planning project,
- The range of results that are expected,
- The success attributes of the cell (those factors against which the team will score or measure alternative designs and procedures),
- The weights assigned to each of the success factors,
- The core cell team(s) and proposed advisers.

Basic preliminary questions to be asked are shown in Figure 11-9. The steering committee will guide the team activities. It is recommended that the core planning teams meet personally (not via e-mail reports) with the steering committee on a weekly basis. The steering committee needs to coordinate total plantwide materials flow logistics and the macro plant layout plans. This ensures that individual cell design teams blend their operations efficiently into the overall material flow of the plant. The steering committee also should coordinate any necessary team member training and cross-training that may be required prior to implementation.

The individual core cell planning teams should be responsible for the analysis and optimization phase of cell design. This is the phase that performs products and parts process reviews and minimizes nonvalue-added activities. Figure 11-10 (page 209) shows the analysis and optimization phase. It should be noted that to correctly define the individual cell location, it will be necessary to develop macro block layouts first. The procedure for this has been described in previous chapters. Those early procedures should take us through steps 1 and 2 of Figure 11-10.

One of the first cell planning tasks is to determine the first cut, macro grouping factors for cell inclusion. The typical commonality or similarity characteristics are:

- Appearance/geometry,
- Components,
- Configuration,
- Customer service/market,
- Degree of quality required,
- Manufacturing operations,

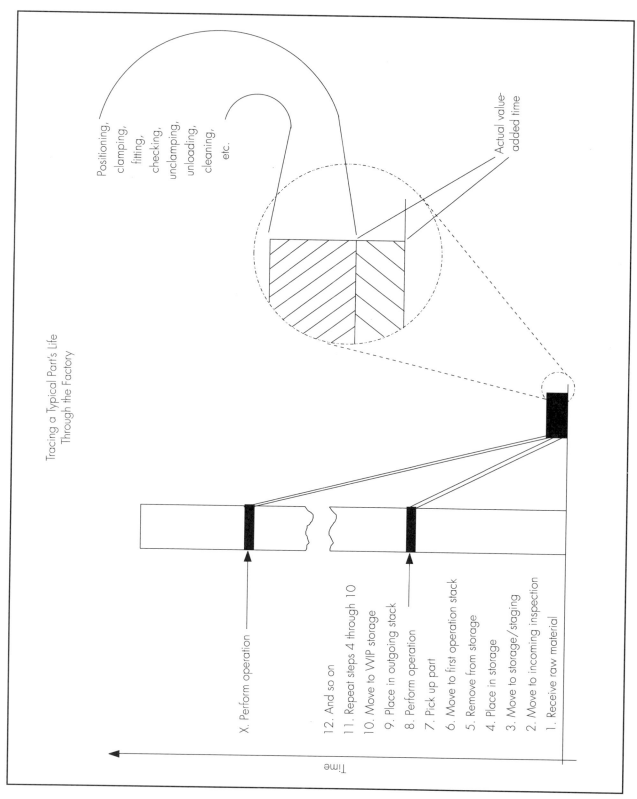

Tracing a Typical Part's Life
Through the Factory

Positioning,
clamping,
fitting,
checking,
unclamping,
unloading,
cleaning,
etc.

Actual value-
added time

X. Perform operation

12. And so on
11. Repeat steps 4 through 10
10. Move to WIP storage
9. Place in outgoing stack
8. Perform operation
7. Pick up part
6. Move to first operation stack
5. Remove from storage
4. Place in storage
3. Move to storage/staging
2. Move to incoming inspection
1. Receive raw material

Time

Figure 11-6. Even what appear to be value-added steps in a production process, when broken down, show nonvalue-added activities that add cost to the product.

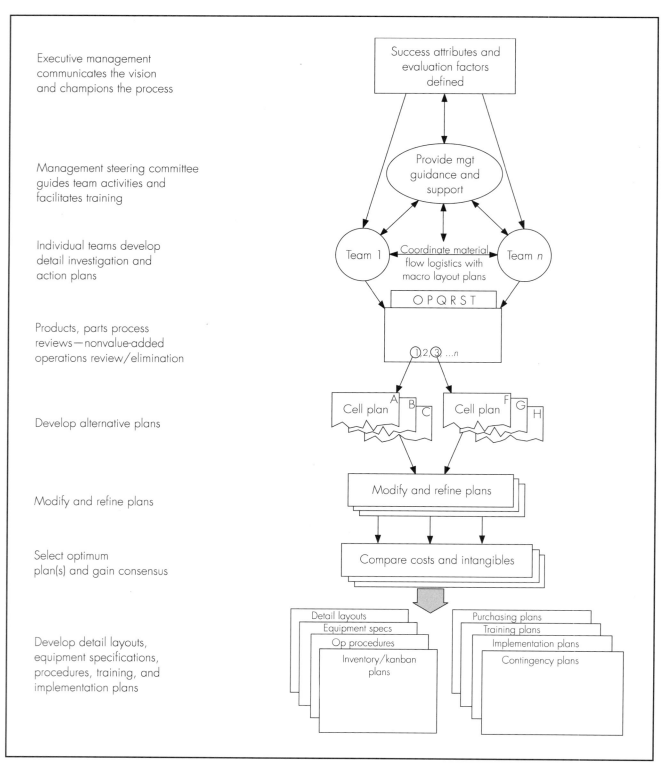

Figure 11-7. *Structure of a manufacturing cell planning effort. Goals and how to measure success lead off the project and are communicated to the enterprise by executive management.*

Area/Function	Core team	Specialists Part-time	Specialists Full-time	Specialists External	Advisory
General management					S
Engineering					S
– Product design		T			
– Product development	T				T
– Tooling/fixturing		T			
– Product research					T
– Manufacturing/process	T		T		
– Industrial		T			
– Analyst/consultant		T		T	
Marketing					S
– Sales					T
– Service					T
Finance:					S
– General accounting		T			
– Cost accounting		T			
Systems					
– Analyst					T
– Process controllers					T
– Applications	T				
Suggestion systems					T
Factory Production					S
– Supervision	T				
– Managers		T			
– Shipping/receiving					T
– Training			T	T	
– Traffic					T
– Shop floor group leaders	T				T
Quality engineering					T
Maintenance		T			
Facilities			T		
Production control					S
– Material	T				
– Purchasing		T			
– Scheduling/planning			T		
– Order entry					T
– Inventory		T			
Personnel/human resources					S
Office administration					T
Cell development	T			T	

Title of table: Suggested Cell Planning Roles/Organization

T = Team representative
S = Steering committee representative

Figure 11-8. *Organization and role definitions should be clearly spelled out by executive management and the cell steering committee. (Adapted from* Making Manufacturing Cells Work; *Society of Manufacturing Engineers; Dearborn, Mich.; 1992.)*

Basic Preliminary Cell Design Questions

- How will we measure success? Can we define the success factors or attributes?
- What weights will be assigned to the success attributes? For example,
  - Is capacity more important than direct labor cost, equal in importance, less important? What weights will be assigned to each attribute?
  - Suppose we can reduce the amount of WIP materials but there may be an increase in indirect materials handling costs. How much of a rise in indirect costs, if any, is allowable?
- Who will be responsible for setting up a training plan and cross-training the operators?
- How will the cell be scheduled? Will a change in the materials control and purchasing procedures be required? Who will be responsible for the scheduling aspect?
- Who will be members of the core cell planning team(s), the management team, the steering committee, etc.? What additional specialists will be required?

Figure 11-9. Far too often, projects are begun before fundamental issues have been resolved. In cell design, answers to these questions will help determine the design approach and where responsibilities lie.

- Number of physical transactions,
- Operation(s) routing/sequence,
- Operation(s) times,
- Profit contribution,
- Quantity produced,
- Raw materials,
- Special/unique handling considerations,
- Special/unique product considerations.

Typical detail grouping characteristics are shown in Tables 11-1 through 11-3.

Note that in the assembly cell category, the customer/market focus is emphasized. The great majority of focused cells are set up based on this one factor alone. This type of focus helps to convince the company's customers that they are truly important to the enterprise. Customers do not feel entirely comfortable seeing their parts or orders mixed with other customer orders in a manufacturing environment. They

much prefer to have a production cell dedicated to their own product. This allows direct communications with the people specifically tasked to complete their orders on the shop floor. Some companies allow their customers to bypass the front office completely for status reports and contact the shop-floor cell team responsible for their work. Normally, status requests are transmitted by telephone, e-mail, or fax directly to a receiving device within the cell.

Table 11-1. Assembly Cells

Commonality or similarity of

1. *Customer/market focus*
2. Product size, weight
3. Number of component parts
4. Number and type of assembly operations and skills
5. Amount of assembly time
6. Testing equipment
7. Heating, cooling, finishing equipment
8. Handling equipment
9. Packaging equipment
10. Quantity per product required

Table 11-2. Sheet Metal Fabrication Cells

Commonality or similarity of

1. Sheet metal alloy
2. Tolerances
3. Size before bending
4. Gage (metal thickness)
5. Number of bends
6. Punched features
7. Finished shape
8. Quantity per part required

Table 11-3. Machining Cells

Commonality or similarity of

1. Raw material
2. Tolerances
3. Size before machining
4. Type of machining operations required
5. Number of machining operations required
6. Clamping or fixturing features
7. Finished shape
8. Quantity per part required

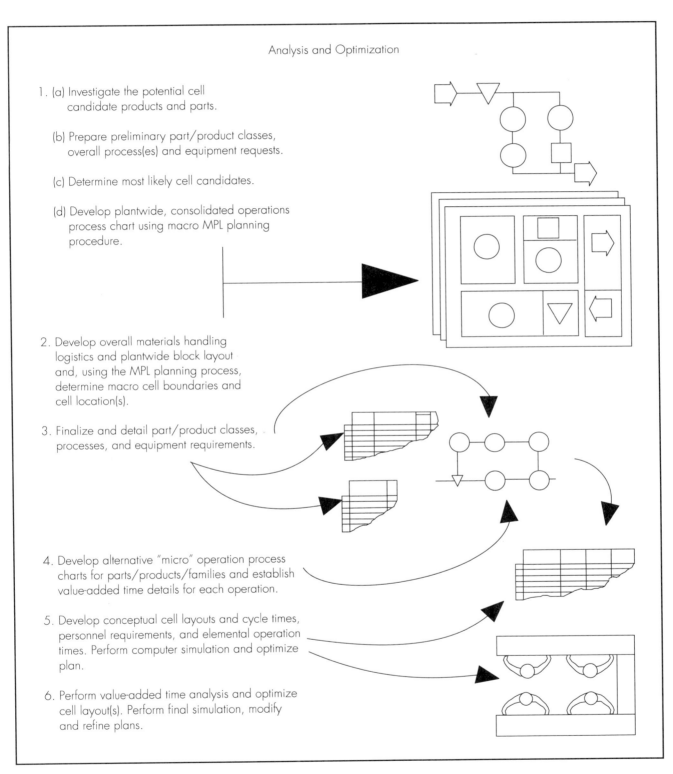

Analysis and Optimization

1. (a) Investigate the potential cell candidate products and parts.

   (b) Prepare preliminary part/product classes, overall process(es) and equipment requests.

   (c) Determine most likely cell candidates.

   (d) Develop plantwide, consolidated operations process chart using macro MPL planning procedure.

2. Develop overall materials handling logistics and plantwide block layout and, using the MPL planning process, determine macro cell boundaries and cell location(s).

3. Finalize and detail part/product classes, processes, and equipment requirements.

4. Develop alternative "micro" operation process charts for parts/products/families and establish value-added time details for each operation.

5. Develop conceptual cell layouts and cycle times, personnel requirements, and elemental operation times. Perform computer simulation and optimize plan.

6. Perform value-added time analysis and optimize cell layout(s). Perform final simulation, modify and refine plans.

Figure 11-10. *In-depth investigation of cell design by core planning teams provides the basis for culling out nonvalue-added activities and optimizing cell design.*

DETAIL CELL PLANNING

Developing detail cell plans and layouts is usually a combined analytical and iterative procedure. One of the first tasks is to finalize the part/product classes, processes, and equipment requirements. This process is capsulized in step 3 of Figure 11-10. Useful for accomplishing this task is a Part/Product Classification Summary Sheet like that shown in Figure 11-11.

This sheet is usually set up in a computer-based spreadsheet format. All of the candidate parts are listed, along with their physical similarity factors and risk of handling damage. The columns headed P1 through P7 are used to record either process sequences or value-added time (or both on two separate sheets). There are also columns for recording and analyzing storage cube requirements. (Storage cube requirements are those point-of-use storage needs normally associated with assembly cells; see Figure 11-12.) For comparing alternative situations, two columns are shown for analyzing the percent of value-added time versus total cell time for individual parts.

Two additional spreadsheets are shown in Figures 11-13 and 11-14, pages 213 and 214. These are for a fabrication cell example and a machining cell example. These charts are normally developed after some preliminary sorts are completed on the part/product classification summary sheet.

Additional sorts on the process routing charts (Figures 11-13 and 11-14) allow development of the most likely cell candidates. These are then segregated into process classes. A further analysis may be performed using the classification summary sheet shown in Figure 11-15, page 215. Similar to our prior analyses of the *P-Q* chart, part/product classes can be further developed using characteristics such as:

• Quantity produced per week, month, year, etc.,
• Average profit margin for the product class,
• Total labor hours for the class.

When your cell classes and cell candidates are finalized, you can use the Process Per Class chart shown in Figure 11-16, page 215. This chart allows us to calculate the number of machines or other pieces of equipment that will be required

in the cell to produce the part quantity requirements. You should be forewarned that this can be a much more complex analysis than the one shown here. For example, if some parts are required on a seasonal or cyclical basis, the analysis is more complex. Some pieces of equipment may be idle for lengthy periods. Conversely, we may have severely understated the same equipment needs in peak periods. Also, setup times must be established based on assumed batch sizes that may change as needs dictate. Similarly, the summary time given per operation and setup is a weighted average for *all* classes of parts. An individual part analysis must be performed if there are wide variances from average values within the classes. A dynamic, computer-based simulation is recommended for complex mixes of part and product classes within a cell.

Once we have established some conceptual cell layouts and cycle times, we then need to optimize the layout and procedures to minimize nonvalue-added time.

Table 11-4 lists some of the typical nonvalue-added activities the planner should be aware of.

Table 11-4. Methods Improvements: What to Look For

Activities to be reduced or eliminated	
• Adjusting	• Pre-positioning
• Bending over	• Positioning
• Carrying or transporting	• Pulling
• Checking	• Pushing
• Choosing	• Reaching
• Cycling to the wrong rhythm	• Resting
• Disassembling	• Searching
• Gaging	• Stooping
• Idle chatter	• Turning around
• Idleness	• Unnecessary paces
• Measuring	• Waiting
• Orienting	• Walking

Figures 11-17 through 11-22 (pages 216 through 219) show a typical before-and-after example of reducing nonvalue-added handling and optimizing an assembly cell layout. All time numbers have built-in allowances for unavoidable delays (personal, fatigue, etc.). Notice the very poor labor balance in Figure 11-18 on page 217. Production was limited to the slowest

Part/Product Classification Summary

Part or product number	Name/ description	Qty per ( )	Similarity factors				Risk of handling damage	Process sequence or VA time (VA)							Number of storage slots/bins	Cube of storage slots/bins	Absolute min cycle time (ΣVA)	% Value-added time $\frac{\Sigma VA}{\Sigma(VA + NVA)} \times 100$	Other comments
			Mkt/build schedule	Shape	Material	Weight		P1	P2	P3	P4	P5	P6	P7					
1.																			
2.																			
3.																			
4.																			
5.																			
6.																			
7.																			
8.																			
9.																			
10.																			
11.																			
12.																			
13.																			
14.																			
15.																			
16.																			
17.																			
18.																			
19.																			
20.																			
21.																			
22.																			
23.																			
24.																			
25.																			
26.																			
27.																			
28.																			
29.																			
30.																			
31.																			
32.																			
33.																			
34.																			
35.																			
36.																			
37.																			
38.																			
39.																			
40.																			

Figure 11-11. *At the detail level, cell planning involves breaking down parts and products into their respective classes, processes, and overall equipment requirements. A matrix such as shown here streamlines this step of the process.*

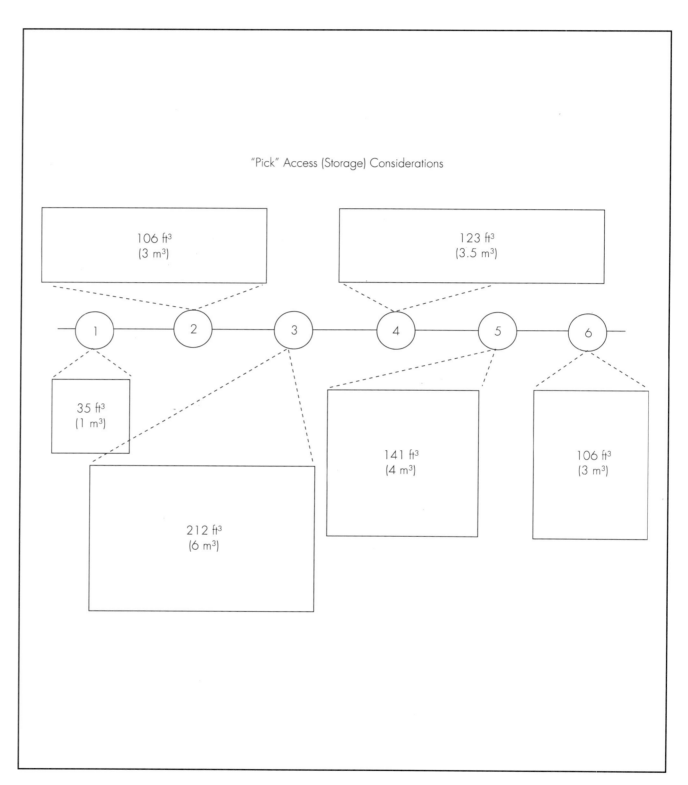

Figure 11-12. *In assembly cells, calculating storage cube needs and allocating the proper amount of storage space and accessibility on the layout are critical to production throughput.*

Classifications by Process in
Developing a Fabrication Cell

Step 1. Develop a spreadsheet comparison chart between machines and components. Use the number 1 to signify that a part is processed on a particular machine. Segregate into like product/part classifications.

	Machine/operation						
Component	Shear	Punch	Notch	Bend	Weld	Totals	Product class
428	1	1		1		3	
493	1	1		1	1	4	
272		1	1	1		3	
231	1	1	1	1	1	5	
222			1			1	
465	1	1			1	3	
243	1	1		1	1	4	
279		1	1	1	1	4	
344	1	1			1	3	
456	1	1	1	1	1	5	
398	1	1		1		3	
561		1	1	1	1	4	
214	1	1		1	1	4	
479	1	1	1	1	1	5	
234		1				1	
Totals	10	14	7	11	10	52	

Step 2. Determine what activities are best left centralized.
In this case all activities may be within the cell (perhaps excluding weld).
Postpone this decision until a detailed time balance is performed.

Step 3. Resort data by product class and probability (most likely) cell candidates.

	Machine/operation							
Component	Shear	Punch	Notch	Bend	Weld	Totals		Product class
231	1	1	1	1	1	5	√√√√	A
456	1	1	1	1	1	5	√√√√	A
479	1	1	1	1	1	5	√√√√	A
493	1	1		1	1	4	√√√	B
243	1	1		1	1	4	√√√	B
214	1	1		1	1	4	√√√	B
279		1	1	1	1	4	√√√	C
561		1	1	1	1	4	√√√	C
428	1	1		1		3	√√	D
398	1	1		1		3	√√	D
465	1	1			1	3	√√	E
344	1	1			1	3	√√	E
272		1	1	1		3	√√	F
222			1			1	√	G
234		1				1	√	H
Totals	10	14	7	11	10	52		

√√√√ Most likely overall cell candidates
√√√ Highest next most likely cell candidates
√√ Lowest next most likely cell candidates
√ Least likely cell candidates

Figure 11-13. *Detail planning for specific functions in a fabrication cell is simplified by use of a spreadsheet format.*

Classifications by Process in
Developing a Machine Shop Cell

Step 1. Develop a spreadsheet comparison chart between machines and components. Use the number 1 to signify that a part is processed on a particular machine. Segregate into like product/part classifications.

| Component | Machine/operation | | | | | Totals |
	Saw	Mill	Lathe	Drill	Pack	
937	1	1			1	3
563	1	1	1	1	1	5
572	1			1	1	3
326	1	1	1	1	1	5
674	1	1			1	3
774	1		1	1	1	4
326	1	1	1		1	4
962	1	1			1	3
737	1	1	1		1	4
581	1	1			1	3
987	1		1	1	1	4
570	1	1	1		1	4
871	1	1	1	1	1	5
Totals	13	10	8	6	13	50

Product class

Step 2. Determine what activities are best left centralized.
*Since every part is sawed and packed, it may be most efficient to centralize these operations rather than splitting them up into cells, e.g., centralize saw at incoming/receiving and centralize packing at shipping.

Step 3. Resort data by product class and probability (most likely) cell candidates.

| Component | Machine/operation | | | Totals | | Product class |
	Mill	Lathe	Drill			
563	1	1	1	3	√√√	A
326	1	1	1	3	√√√	A
871	1	1	1	3	√√√	A
326	1	1		2	√√	B
737	1	1		2	√√	B
570	1	1		2	√√	B
774		1	1	2	√√	C
987		1	1	2	√√	C
937	1			1	√	D
674	1			1	√	D
962	1			1	√	D
581	1			1	√	D
572			1	1	√	E
Totals	10	8	6	24		

√√√ Most likely overall cell candidates
√√ Next most likely cell candidates
√ Least likely cell candidates

Figure 11-14. *The spreadsheet format enables the layout planner to determine which components are best suited for manufacturing cells and which are better left centralized.*

Classification Summary

Product class	Class description	Ω_n	Material handling code	Distinctive or representative feature	Examples of products or parts (part no.)
A					
B					
C					
D					
E					

Ω_n = 1. Quantity produced per week, month, year, etc.
　　　 2. Average profit margin for this product class
　　　 3. Total annual labor hours for this class

Material handling code	Description
a	
b	
c	
d	

Figure 11-15. *Product/part classification can be further detailed in a classification summary such as this that identifies part attributes which could have a bearing on the process type.*

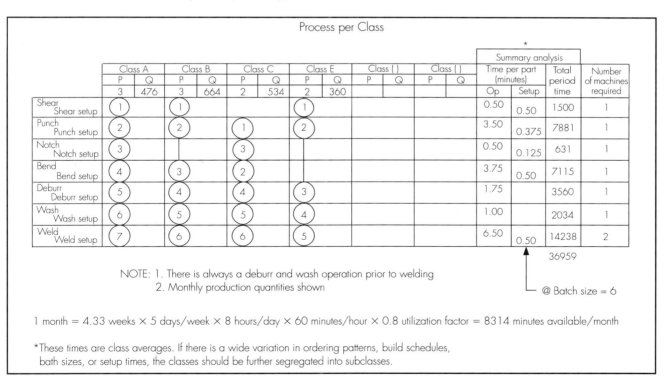

Process per Class

	Class A		Class B		Class C		Class E		Class ( )		Class ( )		Summary analysis Time per part (minutes)		Total period time	Number of machines required
	P	Q	P	Q	P	Q	P	Q	P	Q	P	Q	Op	Setup		
	3	476	3	664	2	534	2	360								
Shear / Shear setup	①1		①1				①1						0.50	0.50	1500	1
Punch / Punch setup	②2		②2		①1		②2						3.50	0.375	7881	1
Notch / Notch setup	③3				③3								0.50	0.125	631	1
Bend / Bend setup	④4		③3		②2								3.75	0.50	7115	1
Deburr / Deburr setup	⑤5		④4		④4		③3						1.75		3560	1
Wash / Wash setup	⑥6		⑤5		⑤5		④4						1.00		2034	1
Weld / Weld setup	⑦7		⑥6		⑥6		⑤5						6.50	0.50	14238	2

36959

@ Batch size = 6

NOTE: 1. There is always a deburr and wash operation prior to welding
　　　 2. Monthly production quantities shown

1 month = 4.33 weeks × 5 days/week × 8 hours/day × 60 minutes/hour × 0.8 utilization factor = 8314 minutes available/month

*These times are class averages. If there is a wide variation in ordering patterns, build schedules, bath sizes, or setup times, the classes should be further segregated into subclasses.

Figure 11-16. *Size of the manufacturing cell and the number and type of equipment pieces needed in it can be calculated from a Process per Class chart.*

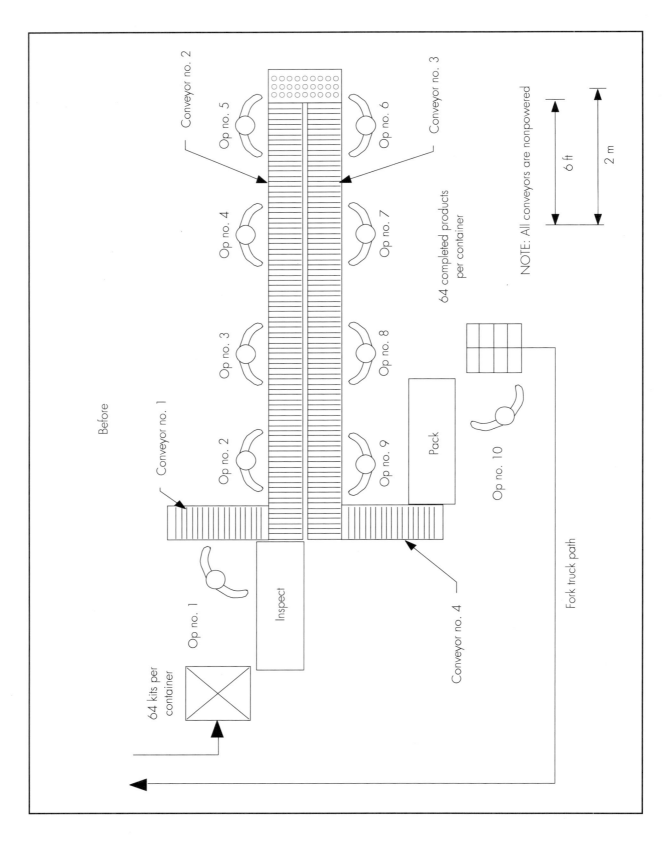

Figure 11-17. *Cell layout showing the "before" situation in a case study aimed at reducing nonvalue-added materials handling and optimizing cell layout.*

Work element	Seconds per unit	Operator balance times	Value-added time
1. Pick up part, place on table	3.0		
2. Inspect	5.0		5.0
3. Push to end of table	1.0		
4. Align on conveyor no. 1	1.8		
5. Walk to incoming container	2.8	13.6	
6. Turn and pull unit to first position	2.5		
7. Perform operation no. 2	7.0		7.0
8. Push to operator no. 3	1.0	10.5	
9. Pull unit into position	1.5		
10. Perform operation no. 3	10.5		10.5
11. Push to operator no. 4	1.0	13.0	
12. Pull unit into position	1.5		
13. Perform operation no. 4	9.0		9.0
14. Push to operator no. 5	1.0	11.5	
15. Pull unit into position	1.5		
16. Perform operation no. 5	7.2		7.2
17. Push on to roller table	1.0	9.7	
18. Pull unit off roller table and position	2.0		
19. Perform operation no. 6	8.5		8.5
20. Push to operator no. 7	1.0	11.5	
21. Pull unit into position	1.5		
22. Perform operation no. 7	7.5		7.5
23. Push to operator no. 8	1.0	10.0	
24. Pull unit into position	1.5		
25. Perform operation no. 8	7.8		7.8
26. Push to operator no.9	1.0	10.3	
27. Pull unit into position	1.5		
28. Perform operation no. 9	8.5		8.5
29. Turn and push unit to conveyor no. 4	2.5	12.5	
30. Pull on to table	1.0		
31. Box makeup	3.5		3.5
32. Pack	4.0		4.0
33. Slide to end of table	1.0		
34. Walk and place in container	1.5		0.5
35. Return to left end of table	2.8	13.8	
Totals =	116.4		79.0
			67.9%

Figure 11-18. In the 35 work elements before analysis, a gross labor imbalance existed. Itemization of operations with attendant times per unit identify the areas needing greatest attention. Operations 30 through 35 form the bottleneck.

Potential Layout Improvements

1. Shorten incoming table. Too much walking (better yet, use certified suppliers and eliminate inspection).
2. Eliminate conveyor no. 1. It is creating unnecessary WIP.
3. Shorten conveyors no. 2 and 3 to reduce WIP and cycle time.
4. Rebalance to reduce the number of workers required.
5. Relayout as an "outward"-looking U-shaped layout.
   (a) Design so people do not face each other (reduces productivity inefficiencies due to nonwork related "chatting").
   (b) Allows for mutual assistance and reduction of slack time.
6. Eliminate conveyor no. 4. It is creating unnecessary WIP.
7. Reduce size of packing table. There is too much walking.
8. Move output of cell next to cell input to reduce forklift truck travel.
9. Consider using smaller containers to reduce WIP (may increase indirect materials handling costs, e.g., number of fork truck trips).

Figure 11-19. An experienced observer's visual analysis of the layout would show these shortcomings.

Work element	Seconds per unit	Operator balance times	Value-added time
1. Pick up part, place on table	3.0		
2. Inspect	5.0		5.0
3. Perform operation no. 2	7.0		7.0
4. Push to next position	0.5		
5. Turn and pack	5.2		4.0
6. Place in container	1.0	21.7	0.5
7. Pull unit into position	0.5		
8. Perform operation no. 3	10.5		10.5
9. Perform operation no. 4	9.0		9.0
10. Push to next operator	0.5		
11. Some box makeup	1.2	21.7	1.2
12. Pull unit into position	0.5		
13. Perform operation no. 5	7.2		7.2
14. Perform operation no. 6	8.5		8.5
15. Perform half of operation no. 7	3.8		3.8
16. Push to next operator	0.5		
17. Some box makeup	1.2	21.7	1.2
18. Pull unit into position	0.5		
19. Perform half of operation no. 7	3.8		3.8
20. Perform operation no. 8	7.8		7.8
21. Perform operation no. 9	8.5		8.5
22. Some box makeup	1.1	21.7	1.1
Totals =	86.8		79.1
			91.1%

NOTE: Box makeup now shared by three operators

Figure 11-20. After the improvements of Figure 11-19, value-added work effort rose to 91.1% of total production time, an increase of more than 23%.

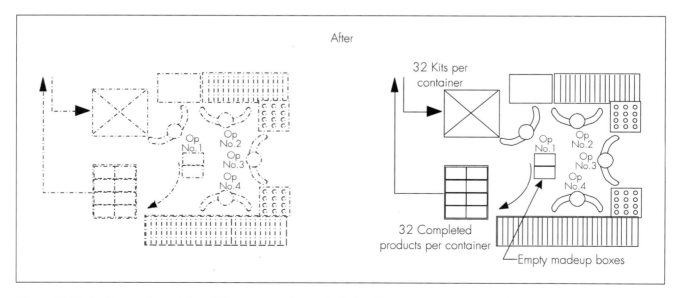

Figure 11-21. In the new layout, the cell has an entirely new look. In this case, output capacity of one 4-worker cell did not quite equal the capacity of the former 10-worker cell, but installing two cells would have resulted in too much capacity.

Before Versus After

Factor	Before	Single "U" cell	Double "U" cell	Improvement
1. Actual production/shift	1250	1128	2256	
2. Theoretical 100% production per shift	2087	1327	2654	
3. Direct labor minutes per unit: Actual Theoretical	3.84 2.30	1.70 1.45	1.70 1.45	D/L costs down 55%
4. Output per worker per shift	125	282	282	225%
5. Cell WIP inventory (units)	114	46*	92*	

*Smaller container size

Figure 11-22. *Results of the case analysis show dramatic improvements in both cost and output per worker. Overall efficiencies (compared to a 100% theoretical maximum) rose from 59.8% to 85.0%.*

operating position (operations 30 through 35). The *after* solution had much higher efficiencies and less time lost to "chatting," etc., but could not quite meet the capacity of the 10-person arrangement. The company opted to run approximately 10% overtime with the single four-person cell rather than installing two cells (which would have given them too much capacity). Lot sizes and WIP inventories were also reduced in this example.

BALANCING LABOR WITHIN CELLS AND ON ASSEMBLY LINES

For most small and medium-sized companies, optimum assembly line labor balances and cell balances are usually achieved by manually or iteratively using a computer-based spreadsheet analysis. Many larger companies, such as those in the automotive industry, have internally-developed computer programs that are an aid in developing balances on long, complex assembly lines.

In most cases, a precedence diagram is required at the start of the analysis. Figure 11-23 shows a typical assembly line example. The arrows show the "precedence required" before an operation can be started. The decimal numbers shown are the labor times required to complete each task. As shown in the figure, operation no. 4 cannot be started until operations nos. 2 and 3 are completed. Likewise, operation no. 9 cannot be started unless operations nos. 8 and 5 are completed. Figure 11-24 (pages 221-223) shows the Ranked Positional Weighted (RPW) method used to determine the number of stations required, the cycle time, and balance. This is an iterative approach that starts in step 1 with a ranked listing of tasks. The position of a task in the descending list is determined by the amount of time required to complete the task *plus the amount of time required to complete all the remaining tasks*. When you assign the first station, you then redo the list and remove that task number from the other tasks that required it as a precedent. You continue to follow that procedure, always selecting the next station to be assigned that has the highest remaining time and no precedents remaining. Finally, when all stations have been assigned, a cycle time is established

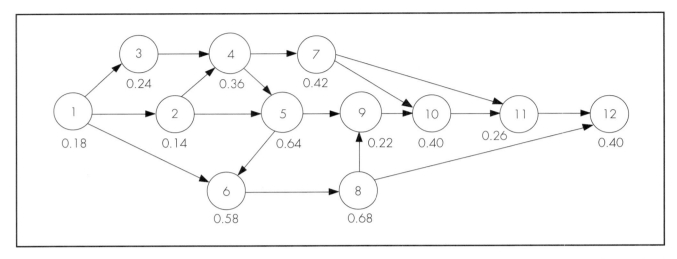

Figure 11-23. *Precedence and timing diagrams such as this plot production flow in terms of what operation must be completed before the next one can be begun.*

that minimizes slack time or imbalance time. In the example shown, a U-shaped cell is being designed so that mutual assistance of several operators may be used to help those stations that do not have enough theoretical time to complete their tasks. As the example shows, with mutual assistance, a cycle time of 0.58 minute would yield a *theoretical* labor efficiency of 97.4%.

In practice, mutual assistance is heavily dependent on the line or cell layout and the teamwork of the cell crew. Successful cells are usually designed in a U shape to allow for mutual assistance. Those operators with available slack time must be positioned properly so they can assist those operators that do not have enough time to complete their tasks (those with negative slack time). Typically, small buffers and small (usually totable) batches are required. Single-piece flow is extremely difficult to achieve on a production line of this type, particularly if more than five or six stations are required. Fatigue factors and other unavoidable factors also have to be taken into account. Table 11-5 lists the typical steps used in trying to achieve a balance.

Table 11-5. Balancing Assembly Cell/Line Operations

1. Prepare a list of operational tasks to be performed, along with the *detailed* elemental time requirements for each task. (Also code the operations for any positive or negative "zoning" requirement.)
2. Prepare a ranked listing of those operations that have the most flexibility with regard to "sequence." Determine the absolute "precedence" relationships and those operations that may be performed simultaneously. Prepare a precedence diagram.
3. Determine which operations are the bottleneck(s).
4. Improve the bottleneck operations (increase throughput via methods and tooling changes).
5. Repeat steps 1 to 4 as many times as is deemed practicable and feasible. (The bottleneck rankings will change at each iteration.) Revise the precedence diagram as necessary.
6. Prepare a ranked listing of the elemental tasks, their individual task times, and their immediate predecessors. Rank one listing in descending order by the sum of all the task times that follow each task *plus* the individual task time (*ranked positional weighted [RPW] method*).
7. Perform analyses using RPW methods, and calculate the labor efficiency.
8. Modify and repeat as necessary.

	Ranked Positional Weighted (RPW) Method (Letting the Process Determine the Most Favorable Balance)			
Step 1	Rank by the sum of all task times that follow in the precedence diagram (indicator of relative importance)			
	Task element	Immediate predecessor tasks	Time to perform	RPW
	1	—	0.18	4.52
	3	1	0.24	4.20
	2	1	0.14	4.10
	4	2, 3	0.36	3.96
	5	2, 4	0.64	3.18
	6	1, 5	0.58	2.54
	8	6	0.68	1.96
	7	4	0.42	1.48
	9	5, 8	0.22	1.28
	10	7, 9	0.40	1.06
	11	7, 10	0.26	0.66
	12	8, 11	0.40	0.40
Step 2	Follow the same steps (no. 3 to completion) as used in the largest candidate method			
	Remove all predecessors from stations which have been assigned and assign next most feasible task (one without predecessors) that meets cycle time requirements (or establish new cycle time)			
	Task element	Immediate predecessor tasks	Time to perform	Station task time
Station A	1	—	0.18	0.42
Station A	3	—	0.24	
	2	1	0.14	
	4	2, 3	0.36	
	5	2, 4	0.64	
	6	1, 5	0.58	
	8	6	0.68	
	7	4	0.42	
	9	5, 8	0.22	
	10	7, 9	0.40	
	11	7, 10	0.26	
	12	8, 11	0.40	

Figure 11-24. *The Ranked Positional Weighted Method tabulates the data from the precedence diagram to ultimately arrive at the minimum number of workstations and cycle times and the optimum balance times.*

		Immediate			
Step 3	Repeat the process				
	Task element	predecessor tasks	Time to perform	Station task time	
Station A	1	—	0.18		
Station A	3	1	0.24	0.42	
Station B	2	—	0.14		
Station B	4	2	0.36	0.50	
	5	2, 4	0.64		
	6	1, 5	0.58		
	8	6	0.68		
	7	4	0.42		
	9	5, 8	0.22		
	10	7, 9	0.40		
	11	7, 10	0.26		
	12	8, 11	0.40		
Step 4	Repeat the process				
	Task element	Immediate predecessor tasks	Time to perform	Station task time	
Station A	1	—	0.18		
Station A	3	—	0.24	0.42	
Station B	2	—	0.14		
Station B	4	—	0.36	0.50	
Station C	5	—	0.64	0.64	
	6	5	0.58		
	8	6	0.68		
	7	—	0.42		
	9	5, 8	0.22		
	10	7, 9	0.40		
	11	7, 10	0.26		
	12	8, 11	0.40		
Step 5	Repeat the process				
	Task element	Immediate predecessor tasks	Time to perform	Station task time	
Station A	1	—	0.18		
Station A	3	—	0.24	0.42	
Station B	2	—	0.14		
Station B	4	—	0.36	0.50	
Station C	5	—	0.64	0.64	
Station D	6	—	0.58	0.58	
	8	6	0.68		
	7	—	0.42		
	9	8	0.22		
	10	7, 9	0.40		
	11	7, 10	0.26		
	12	8, 11	0.40		

Figure 11-24. (Continued)

Step 6	Repeat the process (twice)				
	Task element	Immediate predecessor tasks	Time to perform	Cumulative station task time	
Station A	1	—	0.18		
Station A	3	—	0.24	0.42	
Station B	2	—	0.14		
Station B	4	—	0.36	0.50	
Station C	5	—	0.64	0.64	
Station D	6	—	0.58	0.58	
Station E	8	—	0.68	0.68	
Station F	7	—	0.42		
Station F	9	—	0.22	0.64	
	10	9	0.40		
	11	10	0.26		
	12	11	0.40		
Step 7	Final				
	Task element	Immediate predecessor tasks	Time to perform	Station task time	Slack time
Station A	1	—	0.18		
Station A	3	—	0.24	0.42	0.16
Station B	2	—	0.14		
Station B	4	—	0.36	0.50	0.08
Station C	5	—	0.64	0.64	<0.06>
Station D	6	—	0.58	0.58	0.00
Station E	8	—	0.68	0.68	<0.10>
Station F	7	—	0.42		
Station F	9	—	0.22	0.64	<0.06>
Station G	10	—	0.40		
Station G	11	—	0.26	0.66	<0.08>
Station H	12	—	0.40	0.40	0.18
		Totals =	4.52		0.12

4.52 minutes actual time required × 100

8 stations × 0.58-minute cycle time each

Labor efficiency = 97.4%

NOTE: The 0.58-minute cycle time can only be achieved if we can arrange the layout so that stations C, E, F, and G can be easily assisted during the slack time of the other operations.

Figure 11-24. (Continued)

MULTIFLOOR, MULTISITE SPACE ALLOCATIONS

ANOTHER USE FOR RELATIONSHIP ANALYSES

Clearly, the best condition that we can face is the layout project where all activities are located in one building and on one floor level. In some situations, however, the planner may need to develop layout plans for multilevel buildings. Similarly, the planner also may have a situation where there are several buildings on one site or on several different sites. Given there are important relationships between activities that may be separated vertically or horizontally in two or more buildings, we need a method to allocate activity locations that will provide minimum "pain" and maximum benefit.

The multifloor/multilocation problem is considered by many planners to be much more complex than a standard one-plant layout, and rightly so. In practice, although we usually have more activities to consider and much more work to do, the logic and procedure used remains roughly the same as we have used before.

In this new situation, we are concerned with at least two levels or planes of relationships. One level is concerned with standard relationships that exist on one floor as they do with one-plane layouts. The other plane to be considered is the number of relationships that are satisfied *among* floors or sites.

Figure 12-1 shows a typical listing format for a multifloor/multisite analysis. We will work out a simple example problem which will demonstrate the activity/space allocation work involved.

In our example, we have a two-story building with 23,700 ft² (2,200 m²) available on each of the floors. This gives us a total space availability of 47,400 ft² (4,400 m²). The areas of each activity are shown in Figure 12-1. The combined relationship diagram and spatial relationship diagram have also been completed (shown in Figures 12-2 and 12-3 [pages 227 and 228]). To keep the example simple, we do not place any restraint or constraint on floor loadings, and we assume an infinite floor loading capability. Therefore, any activity can physically be placed on either floor. Likewise, to demonstrate a point and keep the problem as simple as possible, we assume that we *cannot* install any labor-saving devices such as vertical conveyors, elevators or dumbwaiters between floors. All materials must be moved manually.

The only constraints are:

- The president of the company wants to be on the second floor and he has specifically told you so on several occasions.

- Access/egress doors must be on the first floor (since the building is not being built on a hill).

- Due to awkward loads, it is preferred that the assembly area be located on the first floor.

- Given there are people working on both floors, the restrooms must be apportioned so there are restroom facilities on both floors.

Obviously, there is also the space constraint limiting us to placing no more than 23,700 ft² (2,200 m²) on any one floor. What logic should we use

Clay Manufacturing Co.
Example Problem

Activity	Area requirements (ft²/m²)
1. Main employee entrance	—
2. WIP storage area	2,906/270
3. Brake forming and secondary operations	4,844/450
4. Machine shop	2,476/230
5. Quality assurance and engineering	2,153/200
6. Heat treating department	1,830/170
7. Production planning and scheduling	1,830/170
8. President and plant management offices	1,830/170
9. Final assembly department	9,150/850
10. Raw materials storage	3,983/370
11. Incoming materials staging and finished goods storage	5,597/520
12. Punch press department	4,306/400
13. Lunch/break/meeting rooms	2,691/250
14. Restrooms (2)	1,615/150
15. Shipping/receiving doors	—
Total ft²/m² =	45,211/4,200

Figure 12-1. Activity breakdown by function and area requirements for our two-story example facility.

to allocate activities? We already have it. The *logic* is in the relationships.

Theoretically, the best allocation or stacking scheme would sever the fewest number of relationship lines. If we develop several alternative allocation schemes, we can add up the total number of relationship lines that have been severed. For want of a better term, we may call the relationship lines that are severed penalties. If we make our allocations and cut three *A* relationships, plus three *E* relationships, plus five *I* relationships, our penalty score would be the total of all relationship lines cut. In this case: $(3 \times 4) + (3 \times 3) + (5 \times 2) = 31$. That would be a high penalty score for this example. The alternative

with the lowest penalty score is the best allocation. In the example shown, the best score is in the low 20s. Some example allocations are shown in Table 12-1 on page 229.

Obviously, we have tried to keep the example simple. We could have made it more complex by adding more constraints, allowing vertical automation, or by using three or more floors of a smaller area.

Another point to keep in mind is that a similar procedure could be used to allocate facilities across the country instead of within one building. In that case, instead of using relationships, freight distribution costs and optimum delivery timing would be used. I used a similar analytical procedure for an electric power company. The company serviced a large state in the U.S. with almost 100 crew sites scattered throughout its service area. The crew sites were being supplied by 12 regional warehouses. The project involved consolidating the warehouses from 12 to 3 while still maintaining required delivery schedules. An enormous inventory and materials handling cost reduction was achieved. The project allowed the largest reduction in inventories in the company's history.

When problems become more complex it is much more difficult to develop solutions without the aid of a computer. Typical complex examples would include high-rise office buildings such as those used by large banks and insurance companies. One can still develop workable solutions using manual methods coupled with activity-clustering techniques and organizational analysis, but it is far less time-consuming to use a computerized stacking routine for assistance in establishing floor allocations.

COMPUTER ALLOCATION SYSTEMS

Several software companies offer "stacking" analysis routines with their programs. You can find out how much work is involved in these types of analyses by doing the sample problem we just discussed. Computers have helped enormously in completing these types of projects and are recommended for multistory layout projects in particular.

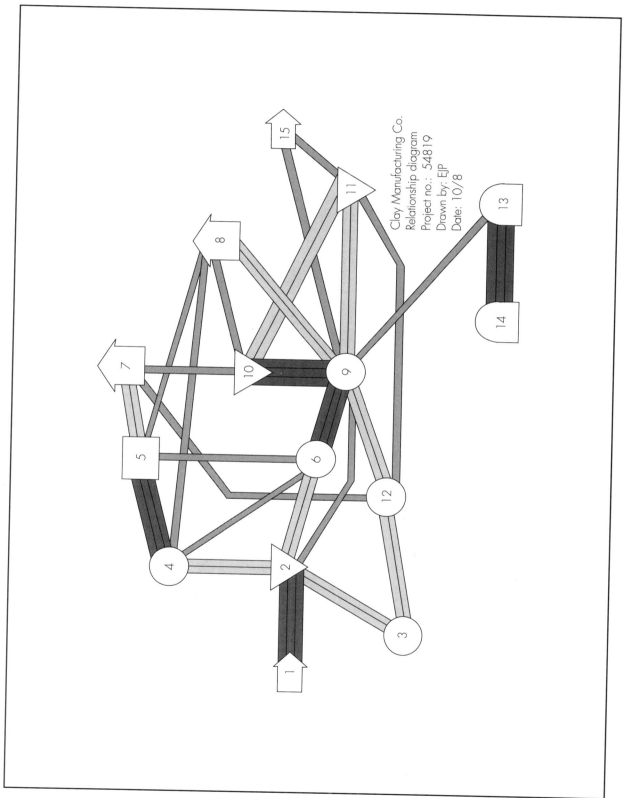

Figure 12-2. Relationship diagram for the example facility, indicating activity areas corresponding to the list in Figure 12-1 and materials handling intensity levels between the areas.

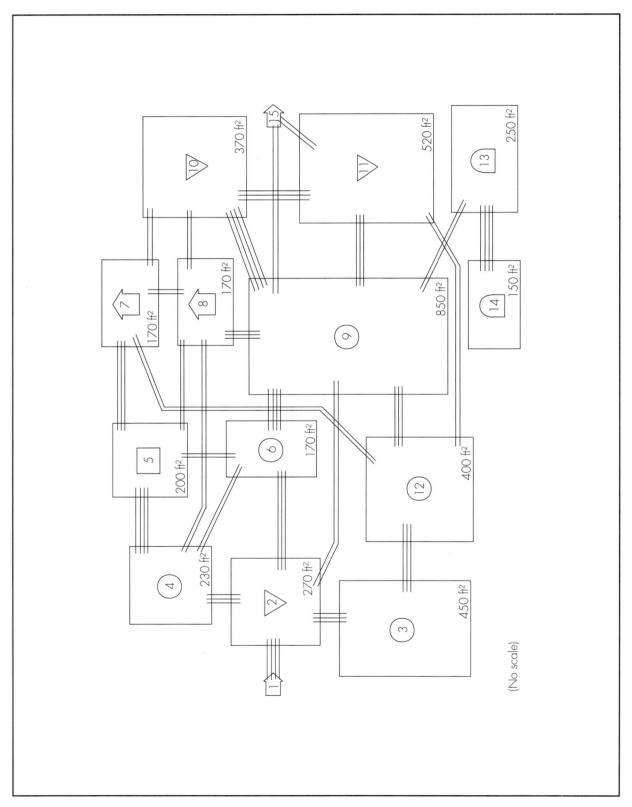

(No scale)

Figure 12-3. The spatial relationship diagram gives the relative sizes of the activity areas to be located on the two floors of the facility. (See Figure 12-1 for activities and area requirements.)

Table 12-1. Clay Manufacturing Co.
Two-floor/Site Worksheet

Activity No.	ft²/m²	Plan A 1st Floor	Plan A 2nd Floor	Plan B 1st Floor	Plan B 2nd Floor	Plan C 1st Floor	Plan C 2nd Floor
1	—						
2	2,906/270	2,906/270			2,906/270	2,906/270	
3	4,844/450		4,844/450		4,844/450	4,844/450	
4	2,476/230		2,476/230		2,476/230		2,476/230
5	2,153/200		2,153/200		2,153/200		2,153/200
6	1,830/170	1,830/170			1,830/170		1,830/170
7	1,830/170		1,830/170		1,830/170		1,830/170
8	1,830/170	1,830/170			1,830/170		1,830/170
9	9,150/850	9,150/850		9,150/850			9,150/850
10	3,983/370		3,983/370	3,983/370			3,983/370
11	5,597/520	5,597/520		5,597/520		5,597/520	
12	4,306/400		4,306/400		4,306/400	4,306/400	
13	2,691/250	1,346/125	1,346/125	2,691/250		2,691/250	
14	1,615/150	807/75	807/75	807/75	807/75	1,615/150	
15	—						
Total		23,466/2,180	21,745/2,020	22,228/2,065	22,982/2,135	21,959/2,040	23,252/2,160

Space available = 23,700 ft²/2,200 m² on each floor

Fixed or requested locations not honored			Plan A	Plan B	Plan C
	A	4	× 1 = 4		× 1 = 4
	E	3			× 1 = 3
	I	2			
	O	1			
Subtotal			4		7

Relationships not honored			Plan A	Plan B	Plan C
	A	4	× 1 = 4	× 2 = 8	
	E	3	× 4 = 12	× 2 = 6	× 5 = 15
	I	2	× 9 = 18	× 4 = 8	× 4 = 8
	O	1			
Grand total			38	22	30

13

COMPUTER-BASED TOOLS

COMPUTER ANALYSES OF MACRO FLOWS AND COSTS

Prior to 1980, it was extremely difficult for small and medium-sized companies to accurately quantify plant layout-related "macro" materials handling costs. The term macro as related to costs usually refers to the materials handling or transport costs between major departments or functions. Macro costs are typically used to compare differential costs between various plant layout alternatives. For the most part, prior to 1980, industrial and manufacturing engineers concentrated their efforts on developing and optimizing "micro" layouts of work centers and individual functional departments. It was an extremely difficult and tedious task to accurately compare materials transport costs between alternatives, let alone attempt to quantitatively optimize an overall plant layout. Most industrial plant layouts were developed using qualitative graphical methods (such as relationship analyses). Many universities experimented with computer-based optimization algorithms, but these were not generally accepted by industry.

Today, many computer-based dynamic simulation programs are available to help the layout planner develop quantitative measurements and optimum layouts. These PC-based programs are used primarily for optimizing work center design. Although they also can be used for designing plantwide logistics and materials handling systems, that is typically not their main focus. During the mid-1980s, other very useful computer-based tools were developed to quantify and optimize materials handling for macro layouts. Such programs have proven very useful to facility planners. They are generally used to establish optimum "rough-cut" macro or block layouts prior to dynamic simulation. In one such program, FactoryFLOW™*, the existing or initial proposed plant layout must first be laid out in a compatible CAD format.

The program computes travel distances and materials transport costs in much the same way that an engineer or accountant would compute these costs and distances by hand if they had an enormous amount of time to do so. But because of the immense speed of high-performance personal computers and engineering workstations, the software can perform these calculations, as well as provide a graphic display of the results, for literally hundreds of parts along hundreds of paths in just a few minutes. The following is a description of the calculation process relative to only one part, an automobile tire, moving along one path. (In the interest of consistency, we track the calculation process using the FactoryFLOW software. This use of Factory-FLOW, however, should not necessarily be construed as an endorsement of that product by either the publisher or the author.)

---

*One of a grouping of several software tools developed by the CIMtechnologies Corporation, Ames, Iowa.

Step One: Compute the Number of Parts Required

FactoryFLOW first multiplies the production quantity by the number of parts per product (taken from the bill of material). This determines how many parts need to be moved between two work areas per unit of time. Thus, if there are four tires per car and the production quantity is 100,000 cars per year, the computer program would determine that 400,000 tires would need to be transported between the work areas every year.

Step Two: Compute the Number of Trips Required

Based on the number of parts (tires in this case) that can be moved on each trip, the program computes the number of trips required to satisfy the production needs. Thus, if 50 tires can be moved with a fork truck on each trip between work area one and work area two, FactoryFLOW would compute that the fork truck would need to make 8,000 trips per year between these two work areas.

Step Three: Compute Path Distances Between the Work Areas

The computer extracts the travel distance from work area one to work area two from the AutoCAD™ drawing. This distance can be specified several ways: it can be entered directly by the user, computed as the straight-line distance between the centers of the two work areas (Euclidean), computed as horizontal and vertical moves between the centers of the two work areas (rectilinear), or measured by tracing the actual path (as drawn by the user) that the device would take to get between the two work areas. Most users allow the computer to calculate the distance automatically, using a rectilinear path or a user-selected path. Because the layout has already been drawn to scale in AutoCAD, all the user is required to do is define (using a mouse or a pen-like pointing device) the pickup and setdown locations on the computer screen. The computer automatically calculates the rectilinear or Euclidean distance between the two points.

Step Four: Compute Materials Handling Equipment Utilization

Once the distance between the two work areas has been determined (let's say 100 ft [30 m] in this case), the program divides the distance by the effectiveness of the materials handling/transport device as specified by the user (or given by the program's default value). Let's say that the fork truck used to move the tires is 50% effective. The low effectiveness value given is equivalent to saying that, after the fork truck delivers the load, it returns to its starting point without a load (deadheads on the return trip, thus it is only 50% effective). Therefore, if the total single move travel distance is 200 effective ft (61 m) per trip, and it takes 8,000 trips per year to satisfy the production rate, the fork truck moves the tires a total of 1,600,000 ft (488,000 m) (303 mi [488 km]) per year (200 ft multiplied by 8,000 trips).

Again, the effectiveness factor is the time that the device travels loaded divided by the total time the device is traveling. Thus, if a fork truck has to drive to a location empty to pick up a load and then drive back with the load, the fork truck is 50% effective. In this case, for every 100 feet of material travel, FactoryFLOW needs to add an additional 100 feet of unproductive travel to account for the total time the materials handling device (a fork truck in this case) is needed.

Next, the computer takes the total move distance and divides it by the default (or user-specified) speed of the device. Thus, if our fork truck travels at an average of 350 ft (107 m) per minute, the travel time for one trip is 0.57 minutes (200 ft divided by 350 ft per minute). FactoryFLOW adds to this time the time needed to pick up and set down a load with this device. Let's say that it takes about 10 seconds to pick up the tires with the fork truck and another 10 seconds to set the tires down. Therefore, it would take 20 seconds, or 0.33 minutes, to pick up and set down tires on each trip (60 seconds per minute divided by 20 seconds to pick up and set down). Our total travel time to move one load of tires from work area one to area two is 0.9 minute (0.57 minutes plus 0.33 minutes).

The total time to move the tires from work area one to work area two each year would be

7,200 minutes or 120 hours (0.9 minutes per trip multiplied by 8,000 trips per year). The computer divides the time the materials handling device is busy by the time it is available to determine its percent of utilization. The software also can factor in multiple servers to determine utilization of a category of materials handling devices.

Step Five: Compute Costs

Finally, FactoryFLOW computes the costs of materials handling by looking at both the fixed and variable costs of the fork truck. In a complex layout, the program calculates costs along each material flowpath for every part or classification of parts specified. (Needless to emphasize, it would be extremely tedious if these calculations were done manually.)

Fixed costs

To compute the fixed costs of the move, the program uses the fixed costs of the equipment and the percentage of operation hours devoted to the move. If the fork truck costs $15,000 new, and it is depreciated on a straight-line method over 5 years, the fixed cost of owning this truck is $3,000 per year whether it is used or not. Now let's say that, in addition to the tires in our examples, this truck is also used to move wheel hubs for a total of 14,400 minutes per year, but beyond moving wheel hubs and tires, it is not needed. Since the truck is needed for the tires 7,200 minutes per year and the wheel hubs 14,400 minutes per year, a third of the yearly fixed costs of the fork truck is apportioned to moving the tires and the remaining two-thirds of the costs to moving wheel hubs. Therefore, the fixed-cost component of the tire movement comes to $1,000 (33.3 percent of tire movement fixed costs multiplied by $3,000 total yearly fixed cost).

Variable costs

The variable cost component of the tire operation is the operator overhead, equipment maintenance, and the energy costs per minute multiplied by the total time needed for the device. (You do not have to worry about collecting all of this data up front: the software contains realistic default values for these times and costs,

and any of these defaults may be changed to exactly match the existing situation.)

In the example case, the tires need 7,200 minutes of fork truck use per year. If the operator's salary and overhead are $18 per hour, the variable operator costs are $2,160 per year ($18 per hour divided by 60 minutes per hour, multiplied by 7,200 minutes of use). If the fork truck uses approximately $2 of propane or electric power per hour, the energy costs of the tire movements are $240 ($2 of energy per hour divided by 60 minutes per hour, multiplied by 7,200 minutes of use). If the fork truck needs approximately 10 cents of maintenance for every hour of use, the total maintenance costs applied to tire movement would be $12 (10 cents of maintenance per hour divided by 60 minutes per hour multiplied by 7,200 minutes of use). Therefore, the total variable costs are $2,412 ($2,160 for labor + $240 for fuel + $12 for maintenance).

For our simple (one-part/one-path) tire-moving example, the total annual costs of moving tires in this layout would be calculated to be approximately $3,412 ($1,000 of fixed costs plus $2,412 of variable costs). Obviously, since the fork truck is highly underutilized, the fixed costs make up a large portion of total costs; however, the ratio of fixed to variable costs varies greatly between the several types of equipment typically found in an industrial facility. Therefore, it is very important to study both of these components (fixed and variable) when working to arrive at an optimal materials handling system and layout.

The great benefit in using a focused computer program is that, once the original production data and pickup and setdown points have been established, the program can calculate the total plant's materials transport costs very quickly. It therefore allows the user to make layout changes in the computer and calculate the total cost effect of the layout changes very quickly. This would be an almost impossible (and most boring) task if one had to calculate total costs manually. Because the manual calculation task is so tedious, it seldom gets done without the aid of a computerized system. Prior to the advent of software programs, optimum plant layout solutions were usually developed as qualitative "best guesses."

DYNAMIC COMPUTER SIMULATION

Well-designed manufacturing operations, cells, or assembly lines will have reduced idle time or slack time to an absolute minimum. Unfortunately, there are no general rules to determine how many analysis iterations or alternatives will be required to reach an optimum condition. Although academia has tried since the 1960s to develop useful quantitative queuing, sequencing, and simulation models for manufacturing, they have been only partially successful. With the ever-increasing power of desktop computers, however, commercially-available dynamic simulation models have proven to be one of the most effective analysis aids for designing manufacturing plants.

A history of computer modeling by manufacturers of discrete products can be traced back to the late 1950s and 1960s. During that period, large companies like General Motors and Boeing first attempted optimization analyses on the mainframe computers of that era. Much of their detailed work, however, was diverted to the numerical-controlled machine tool revolution and the computer numerical controlled equipment we now take for granted. Today's manufacturing simulation programs were taken over and nurtured by academia in the 1970s and 1980s, but have since been commercialized and moved out of the university environment. They have been put to effective use in many real-world situations and have definitely earned a place in the professional facility planner's toolbox.

Total real-world conditions and variables may still be a bit too complex to allow the facilities planner to depend totally on a computer simulation, but the simulation can definitely help. Dynamic simulations can help you develop layout alternatives, project expected throughput, determine the effects of equipment breakdowns, size buffer storage requirements, and determine equipment requirements and utilization. A listing of some of the widely known simulation software vendors is included in Table 13-1.

Simulation software will help the planner determine the effects and optimum size of buffer storage between operations. Simulation will also show where the production bottlenecks are located and how they may change when alterations are made to the cell or line layout. The use of simulation software will help determine what happens when various pieces of equipment break down—where and how fast products will accumulate until repairs are made. In this regard, actual distribution curves depicting the historical or projected *frequency* of equipment downtime or random events can be simulated. This type of analysis is far better than using "average" equipment downtime statistics. The other two main outputs of a computer simulation model are the expected production throughput for various scenarios and equipment utilization (both production equipment and materials handling equipment).

Without the benefit of a dynamic computer simulation, how does the planner develop his factory layout, determine the number of machines or pieces of equipment required, establish buffer storage area requirements, etc.? Typically a static capacity planning approach is used. That is to say, *average* production requirements, adjusted for peak conditions as appropriate, are used. These average or peak figures can help determine the maximum *static* throughput of a particular piece of equipment or a particular human operation, e.g., a bottleneck condition. However, using a capacity planning approach alone makes it difficult, if not impossible, to determine in a reasonable amount of time the effect of interacting influences.

For example, what is the benefit of a particular machine that can easily handle an average production rate of 60 pieces per hour, if the sporadic variances of downstream production will periodically *starve* the machine of any input? Once starved, the planned daily production output will not be met. In a complex production environment, a seat-of-the-pants approach to increasing throughput might be to add more equipment or operators because the machine is not achieving the 60-piece-per-hour rate. This would be exactly the wrong thing to do. A dynamic simulation would have pinpointed the actual downstream *system* problem.

One might say that he or she would not need a computer and software system to find the problem; all that would have been needed was a good industrial engineering delay study. Such a study would indeed pinpoint the parts starvation prob-

Table 13-1. A Sampling of Simulation Packages and Vendors

Simulation Packages and Vendors	
Software	Company
1. ARENA™	Systems Modeling Corp. Sewickley (Pittsburgh), PA
2. SIMFACTORY™	CACI Products Company La Jolla, CA
3. PRO MOD™	PMC of Utah Orem, UT
4. MICRO SAINT™	Micro Analysis & Design Boulder, CO
5. SIM KIT™	IntelliCorp Mountain View, CA
6. GPSS/H™	Wolverine Software Corp. Annandale, VA
7. SLAM II w/TESS™	Pritsker Corporation Indianapolis, IN
8. AUTO MOD II™	Auto Simulations, Inc. Bountiful, UT
9. WITNESS™	AT&T ISTEL Cleveland, OH
10. MIC-SIM™	Integrated Systems Tech. Knoxville, TN
11. TAYLOR II™	F&H Simulation, Inc. Orem, UT

lem, but it is doubtful that it would define the total *system* problem. Performing a delay study at each position and given enough study and analysis time, the industrial engineering (IE) study could eventually determine the *system* problems. Keep in mind, however, that this type of IE study is typically a reactionary, *after-the-fact approach*. We are talking about designing a workable system on paper first, *before actually implementing* the production system. That is the true benefit of simulation—correcting potential problems in the planning and design phase, not

after a production system has been placed on the shop floor.

Earlier we touched briefly on queuing models. Generally, queuing models are considered to be analytical or purely mathematical techniques. They tend to work with mean values of inputs and constraints. Although pure simulation language programs are available, the most popular simulation models today generally employ symbolic or visually animated techniques. Valid dynamic simulation models do have to have analytical queuing models as part of their makeup. However, in addition to the capability of working with mean values for parameters, most simulation models have statistical analysis routines, which allow working with random processes or projected parameter distributions. Several more comprehensive models have complete production scheduling capabilities as well.

Before discussing the categories of simulation software, it is important to recognize that simulation *is not* an optimizing tool. That is to say, if you input all of your data and constraints, a simulation model will not go through its calculations and automatically give you the optimum layout for your production line or plant. It will only inform you of the results of what you have modeled. It is still left to the planner to change the layout or other inputs, rerun the model, and get new results. There is no limit to the number of changes one can make, nor is there any guarantee that you will ever achieve the optimum. You still need some experience or intuitive reasoning to make each successive change better than the last one modeled.

Basically, there are three major categories of simulation software: *programming language, simulation language,* and *simulator animation*.

The programming language category generally includes the least expensive software packages. This software is used extensively in universities and tends to be the hardest to learn and the most difficult to use, particularly for the nonprogrammer. It is, however, the most flexible and portable type of modeling system. It generally requires some programming experience to use effectively. The output from this type of program is difficult to understand, especially for most nonprogrammers and manufacturing managers. Therefore it should be enhanced extensively for management presentations.

The discrete-event *simulation language* offers flexibility and is probably the most comprehensive category of simulation software available today. Simulation language software uses a set of standard mathematical building-block constructs to put a model together. "Standard" refers to relatively easily learned standards within each of the packages themselves. It does not necessarily imply that the various packages will talk to one another or exchange files between competing programs. The better-known software packages in this category include graphic animation output. Most of these programs are relatively complex but easier to learn than those in the programming language category. Programs in the simulation language category are more flexible than the animated "simulator" versions, but as computing power increases, the differences between simulation language packages and simulator packages may become blurred.

In a simulator or animation-type of simulation program, higher level constructs or building blocks are used. These may be considered "black boxes." Each block or box contains a complex mathematical function or tool that the model builder does not necessarily have to see. For example, some blocks may be focused on a particular manufacturing technique, such as machining or assembly. One could compare these to a mechanical erector set or Lego™ set, although, admittedly, that is probably an oversimplification. These programs are the easiest to learn and the most user friendly of the simulation programs, but they lack the flexibility of the language programs.

For developing new plant configurations, the most useful (and most impressive) simulation programs include true three-dimensional layout viewing capabilities. The best programs have the ability to view the plant floor as well as elevated equipment from several perspectives. These types of programs are extremely useful for high-level management presentations and have a very high visual impact. Graphic presentations are usually much easier to sell to upper management than pages of text-based statistics. Many simulation programs have visual graphic capabilities, but only a few have true, to scale, graphic representations. Some even have the ability to import existing CAD-based drawings.

Because of the cost of simulation software and the time it takes to learn and master a program, the great majority of production simulations have been done only in large companies. Larger companies almost always have projects under way that can benefit from a computer simulation analysis. They also have the resources to devote full-time staff to simulation efforts. However, as more competition in simulation develops, software prices may well drop to a more affordable level for the smaller manufacturer. In any event, smaller firms can still achieve the benefits of computer simulation by retaining outside consulting services.

Before embarking on a simulation project, you should recognize that developing a typical model and validating it is not a process that can normally be done in 1 or 2 weeks. It is not uncommon for the development and validation of a single cell model to take 4 to 6 weeks of work effort. Complex models can take significantly longer. It could easily take 3 to 5 months of a *team effort* to develop a complete simulation and analysis for a plant producing two dozen or so different products. After developing the functional specification, the development of the model itself will probably take 7 to 9 weeks. Following validation of the model, the analysis will typically take 2 to 4 weeks more. These very gross estimates assume (1) you already have a preliminary gross layout from which to start, and (2) the simulation user has been trained or a simulation consultant is employed before starting the development of the model. On the positive side, you will learn more about the company's operations in that 3- to 5-month time period than you would have learned in 5 years without the simulation.

DYNAMIC MODEL BUILDING*

Computer simulation is a powerful tool for evaluating the performance of manufacturing processes. Simulation provides performance data including throughput, cycle time, total ship-

*This section contributed by Jerry W. Hoskins, owner, Manufacturing Engineering, Inc., Columbus, Ohio.

ments, cost, and several other indices. It will also show bottlenecks in the process and allow the designer to evaluate alternative solutions.

In the past, many middle managers and planners regarded simulation as a complex and cumbersome process that added considerable cost; however, in the last several years simulation software packages have become very user friendly. Today, packages are available that require no special programming, as the process times and logic are entered through pop-up menus or tables. Ideally, every process design engineer should have access to a simulation package and be trained to use it. It will save both money and time in the design and evaluation process.

Another major benefit to the engineer is that simulation tools provide an excellent medium for communicating the design results. A good simulation package will have an excellent animation capability, and some even have three-dimensional (3D) capability that can be initiated with no special drawing requirements. In addition, your simulation package should have the ability to produce color charts and graphs that summarize your results. These tools serve the engineer well when communicating his or her design to the management team, as the model literally provides a storybook of pictures backed up by the process data. It is not uncommon to see a management team in awe at seeing the design perform right before their eyes, and rightfully so. Communication is as much a responsibility of the process designer as doing the actual design work. This tool makes it easy.

As an example, one simulation tool that provides a significant amount of capability (including 3D animation) and is easy to use is Taylor II™ by F&H Simulations. We will use this tool to illustrate the use of simulation to analyze an electro-optical assembly process. (Again, using this particular tool for illustration does not necessarily constitute endorsement by the publisher or the author.)

The cell is shown in Figure 13-1 and consists of two identical assembly cells supplied from a materials receiving area. The model was constructed as part of the design process. The objective was to design a cell that would assemble 250 of these electro-optical devices within the

first 6 months of operation. It also was designed to have the ability to double the capacity to 500 units.

The model has the ability to run either the one cell for lower production volumes or both cells for the high-rate production. This model provided for significant inputs during the design process and some of the key things we learned are summarized below.

- Addition of the tables (buffers) at no. 9 and no. 23 increased the throughput.
- Because two types of skilled workers are required to assemble the unit at no. 11, the model showed a significant performance increase by adding an additional assembly station (no. 14) without adding any more people. This is because only one worker can realistically work on the unit during assembly.
- It was determined that overtime was required the last few weeks to meet the production shipments.
- Supplier delivery requirements were determined by using this model.
- Worker no. 28 performing the kiting operation is underutilized.

Figures 13-2 and 13-3 show important performance measures of the assembly cell. Any data tracked through the model can be analyzed and graphed in this manner.

The simulation tool has a very strong advantage in dealing with problems that have variability — in other words, things that change from piece to piece or over time. Most planners and engineers have historically been locked into doing analyses based on "averages," and that is a relatively good first approximation to what happens *on average*. However, this is inadequate for determining queue sizes, response times, and many other parameters. Simulation provides an easy and effective manner for handling these variations.

To illustrate, it takes 129 minutes to assemble the optics of item no. 7 in Figure 13-1, yet we know there is considerable time variation from one assembly to the next caused by both mechanical and human variability. In this case, we estimated the variability with a normal distribution using the average of 129 minutes and a standard deviation of 20% (26 minutes). Taylor II allows this variability to be easily added by

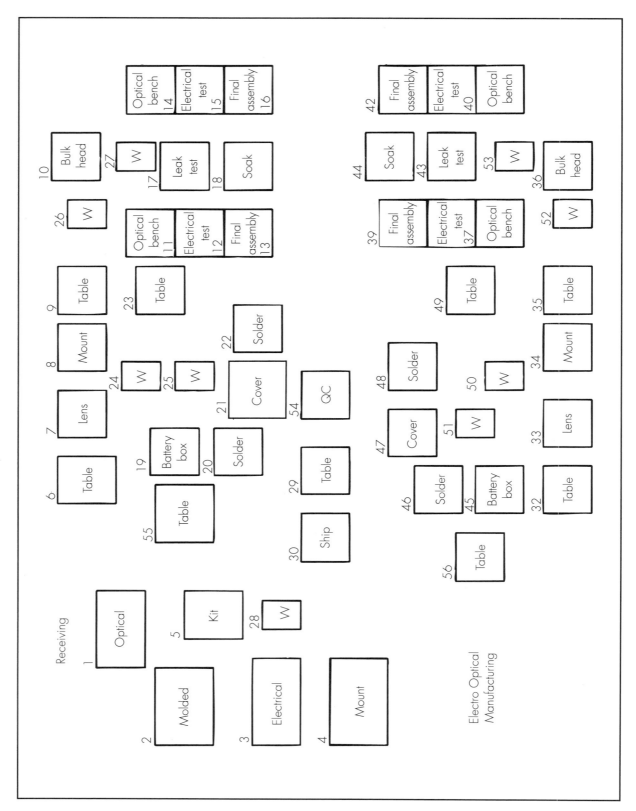

Figure 13-1. *Computer simulation tools enable the planner to quickly rearrange work stations, such as the manufacturing cells here, to achieve optimum output per worker from the layout.*

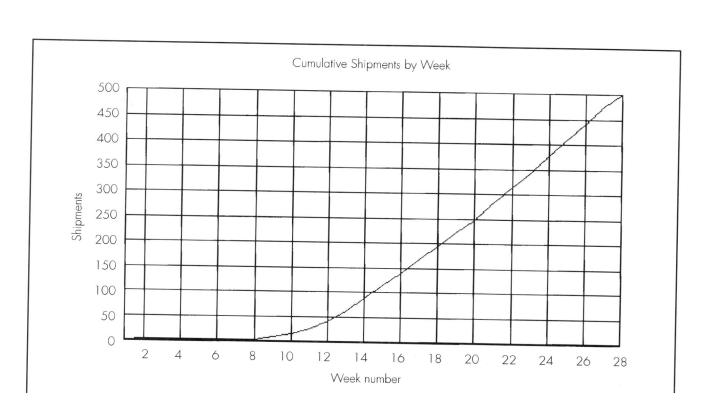

Figure 13-2. Simulation model software can also produce performance measurements such as this, based on the cell layout.

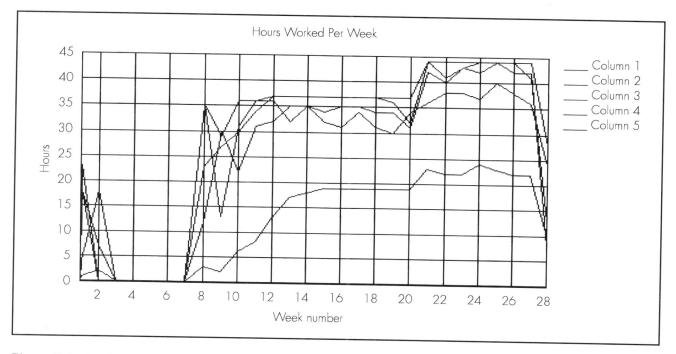

Figure 13-3. Another performance measure generated by the computer simulation model software. Comparative performance metrics can easily be calculated through use of computer-generated productivity statistics.

selecting "Normal Distribution" from a list in a menu. Adding this level of variability to the model has a dramatic effect on the model per-formance, which is easily adjusted by adding buffers and making other changes to the layout. We have found it more important to model the variability than to spend time worrying about the exact amount of variability that exists. If you are worried about the impact on performance, change the standard deviation to 10% and re-run the model. (Planners often become bogged down in simulation by consuming large amounts of time getting the data which often can be easily obtained to the accuracy needed from people on the production floor or from process experts.)

Now that we have designed the assembly cell and have the necessary performance measures analyzed and charted, we are ready to make the presentation to management or to our prospec-tive customer. At this time a 3D simulation is extremely useful. In using Taylor II, we simply select the 3D option and the model appears in 3D form. Several choices also can be made for machines, tables, etc. In this case, the standard defaults were selected, with the resultant model displayed in Figure 13-4.

SECOND EXAMPLE

To further illustrate, we will utilize simulation to redesign an existing manufacturing process of a transducer manufacturer. This manufacturer wanted to improve the time from order receipt to shipment. Its current performance was about 5 weeks turnaround for most products, which forces the company to stock inventory to pro-vide rapid deliveries. This is expensive and, over time, results in high inventories of slow-mov-ing items and out-of-stock situations for fast-moving items.

An assessment of the manufacturing process pointed out three major impediments to achiev-ing reduced cycle time.

1. Production was accomplished in lots of 10 to 20 items, rather than one. This slows throughput because the process resources are blocked until a whole batch is completed.

The inevitable outcome is increased WIP and increased finished-goods inventory.

2. The product was moved around the plant to equipment areas, rather than having the nec-essary equipment in the production line. This functional layout results in lost time, increased WIP and increased labor cost.

3. The manufacturing process was not de-signed to efficiently process the large prod-uct variations. There were bottlenecks and a need for additional process technology.

The process was redesigned to handle one-piece flow. This required both a change in cul-ture and in the manufacturing process itself. For example, the 4-hour curing oven that mandated large batch sizes was replaced with a continu-ous belt type of oven. The belt oven provided continuous flow throughout the process and de-creased the process residence time. The modifi-cations provided an opportunity to establish the manufacturing cell shown in Figure 13-5.

The 3D version of this model is shown in Fig-ure 13-6. This cell has all the equipment and pro-cesses necessary to produce the transducer: welding, assembly, electrical test, weight test-ing, temperature compensation, and shipping.

This revamped process model shows a reduc-tion in cycle time from 20 days to 4 days, and an increase in net income of $2.3 million over 5 years. The manufacturing cell includes *kanban* technology, which was very nicely simulated with the Taylor II package.

The animation provided with simulation is an outstanding training tool. People can see what is supposed to happen, and it is much easier to relate to a dynamic model than a CAD drawing or other visual aid. This is because one can "see" the product moving through the various pro-cesses within the model. These modeling tools are also useful for modeling conveyors, ware-houses, and automatic guided vehicles (AGVs).

Simulation will eventually become as com-mon a design tool for the engineer and layout planner as the spreadsheet. It is a tool used to get at the *real* design issues in a process charac-terized by large amounts of variability. Simula-tion dramatically reduces the risk of failure and provides a marvelous medium through which to communicate design concepts.

Figure 13-4. A presentation-quality 3D model is automatically generated by the software, making senior-management buy-in an easier hurdle to overcome.

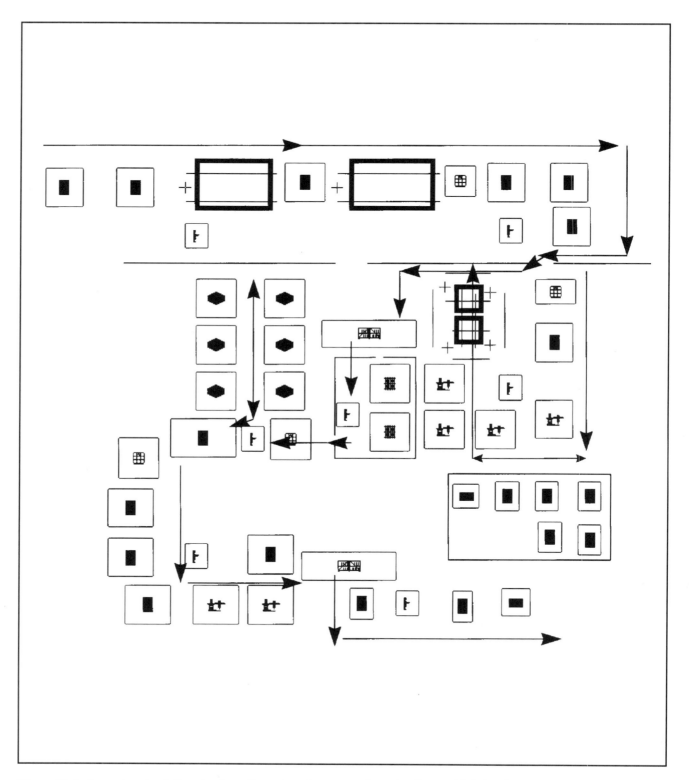

Figure 13-5. Computer simulations to streamline a transducer manufacturing line enabled the planner to redesign the process to handle one-piece continuous flow in the layout shown here.

Figure 13-6. The 3D version of the transducer manufacturing model. Through this simulation, dramatic reductions in cycle times and substantial increases in revenues were shown.

14

EVALUATING ALTERNATIVES

TANGIBLE AND INTANGIBLE DECISION FACTORS

Tangible cost comparisons between layout alternatives are generally made using discounted cash-flow techniques. Some companies use simple payback methods and still others use an internal rate-of-return "hurdle" method. Construction or rearrangement costs are relatively easy to estimate but are not the only consideration that should be used in evaluating alternatives. Many intangible factors are usually involved when deciding between alternatives. Even when alternative implementation costs are not comparable by 20% or more, other intangible factors may be even more important.

In addition to tangible implementation costs, typical intangible factors that may need to be evaluated and scored for each alternative are:

- Risk of lost production (downtime, schedule, etc.);
- Flow of materials effectiveness;
- Fit with the corporate manufacturing philosophy, e.g., focused product cells versus batch manufacturing process layout, meeting space and inventory objectives, reduction of manufacturing lead times, etc.;
- Flexibility for expansion (or contraction);
- Flexibility for change or rearrangement for new product lines;
- Safety, human factors, ergonomics;
- Ease of personnel movement and communication;
- Ease of supervision;
- Appearance or meeting corporate image goals, state-of-the-art installation, customer perceptions, etc.;
- Security;
- "Fit" with the long-term master site, building plan;
- Ease of implementation;
- Ease of housekeeping;
- Effect on employee morale;
- Technical (maintenance) resource capabilities, availability, requirements.

Additional factors unique to each company can be added to this list.

SETTING THE FRAMEWORK FOR TEAM DECISION-MAKING — USING WEIGHTED FACTOR EVALUATION

The factors that will be used in the analysis are usually chosen by the top management team. Ideally, they will include the *success attributes* that were originally developed at the start of the planning process as well as additional factors. It is wise to have top management assign weights to each evaluation factor or attribute. Operating management or the persons who will be responsible for actual departmental operations should then be asked to score each layout based on the factors given.

It is not necessary for the operating managers to initially know the weights assigned to each factor. It is better that these weights be revealed after scoring each alternative plant layout. Using this approach will help eliminate bias in the scoring of the alternatives.

Weighted factor analyses are usually used in evaluating the intangibles and even in evaluating costs in some instances. Figure 14-1 shows a typical rating sheet.

A vowel/letter rating system or a simple one-to-five or one-to-ten rating scale can be used by the operating managers to judge each plant layout on how well it achieves the attribute or factor listed. Typical layouts score high on some attributes and low on others.

After the attributes are scored, they are multiplied by the original weights assigned to each attribute by top management. The weighted scores are then totaled.

A typical scored analysis sheet is shown in Figure 14-2.

Your own judgment must be used in evaluating the totaled scores. Some planners say that a 25% difference in the total score between alternatives is significant; others say that a 15% difference is significant. Whatever score differential you decide to use for your project, it is important to understand that the weighted factor analysis is ideally the culmination of a joint effort. By this point in time, many interested supervisors and managers concerned with the layout have participated in establishing the logic (flow, relationships, etc.) and the evaluation.

The layout will then truly be a team effort which will have several champions in the organization. When there is considerable support for a particular plan organization, it is almost sure to succeed. The important point is to get others involved in the project from the start, and keep them involved. The layout logic will speak for itself and should eliminate most of the normal bickering found in plant layout projects that do not follow an established procedure.

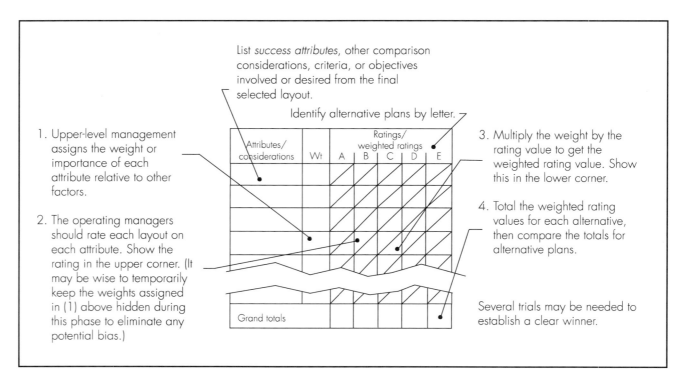

Figure 14-1. *Top management's selection of the analysis factors for comparing layout alternatives and assigning weights to those factors provide the inputs for creating a weighted factor evaluation using a chart such as this.*

Weighted Factor Analysis

Plant/area _____ Many Foods, Ltd. _____ Project ___ FOODS 52101 ___ Date __ 2/12 __

Alternatives: description or name

A. __Alternative #1 East side_____

B. _____

C. __Alternative #2 West side_____

D. _____

E. __Alternative #3 North side_____

Weight set by ____Senior staff consensus____ Ratings by _____

Attributes/ considerations	Wt.	Ratings/weighted ratings					Comments
		A	B	C	D	E	
Initial construction costs	9*	8 / 72		9 / 81		7 / 63	
Efficiency of operations (ongoing costs)	9	9 / 81		9 / 81		8 / 72	
"Togetherness" and communications	6	8 / 48		7 / 42		7 / 42	
Flexibility	8	8 / 64		9 / 72		8 / 64	
Ease of future expansion	7	7 / 49		9 / 63		7 / 49	
Product quality and freshness	10	7 / 70		9 / 90		8 / 80	
Sharing of people and equipment resources	4	4 / 16		4 / 16		5 / 20	
Risk/production loss	10	4 / 40		8 / 80		7 / 70	
Associate convenience	6	5 / 30		5 / 30		5 / 30	
Safety of access and egress	3	9 / 27		9 / 27		8 / 24	
Corporate image/ aesthetics	7	7 / 49		7 / 49		8 / 56	
Totals		546		631		570	

*Weight for initial construction cost is shown for informational purposes only. Estimated actual costs will be used for quantitative comparisons between alternative plans.

— The winner

Figure 14-2. *A completed Weighted Factor Analysis form. The importance of evaluating alternatives using this method is that it is the result of a joint effort, with all members of the team having participated.*

15

CLOSING NOTES ON RELATIONSHIPS AND MATERIALS HANDLING EQUIPMENT CHOICES

BENEFITS OF A WELL-DESIGNED MATERIALS HANDLING SYSTEM

The best-designed plant layout cannot overcome a poor choice of materials handling methods or equipment. Many practitioners believe that the selected materials handling process, coupled with proper space allocations, is *key* to the plant layout process. Simply basing activity relationship importance (and therefore closeness) on the number of connecting transactions can be very misleading. One cannot separate materials handling analysis from the plant layout process. Each segment depends on the other.

For example, suppose we have a plant 600 ft (183 m) long by 100 ft (30 m) wide. Let's also suppose that the only way the plant can expand is in the long dimension, say to 700 ft (213 m) long by 100 ft wide (see Figure 15-1(a)). Most materials are on pallets and 60% pass immediately from receiving into the first operation (storage); therefore there are a high number of transactions. The remaining materials, accounting for 40% of transactions, move directly on pallets from receiving to the second operation. Let's also assume that the second operation after the receiving function is expected to need a great amount of additional space during the next expansion. All of the other areas are expected to remain relatively constant in size. Forklift trucks are selected as the prime mode of materials handling.

If we use a standard from-to chart and relationship analysis, we would place the first operation immediately adjacent to the receiving area. This would be considered an *A* relationship. The second operation would probably be placed next in line to the first operation; this, likely, would be considered an *E* relationship. However, if we locate the second operation as shown, the next expansion would force us to *relocate every department,* (with the exception of receiving and the first operation) to allow for an expansion of the second operation. Suppose instead that we place the second operation at the far end of the plant, near an expandable wall, as shown in Figure 15-1(b). We then install a powered conveyor system, automatic guided vehicle system, drag-line cart system, or other "smart" single-tracked system for materials handling between receiving and the second operation. This would not only allow a relatively easy future expansion for the second operation, but provide a potential materials handling savings as compared to using fork truck travel between receiving and all operations.

This simple example shows why activity relationships and transactions alone can not and should not be the final determinant of an optimum plant layout. Using these factors alone, without regard to materials handling methods/costs, expansion flexibility, and other dominant physical constraints can actually result in a very poor plant layout.

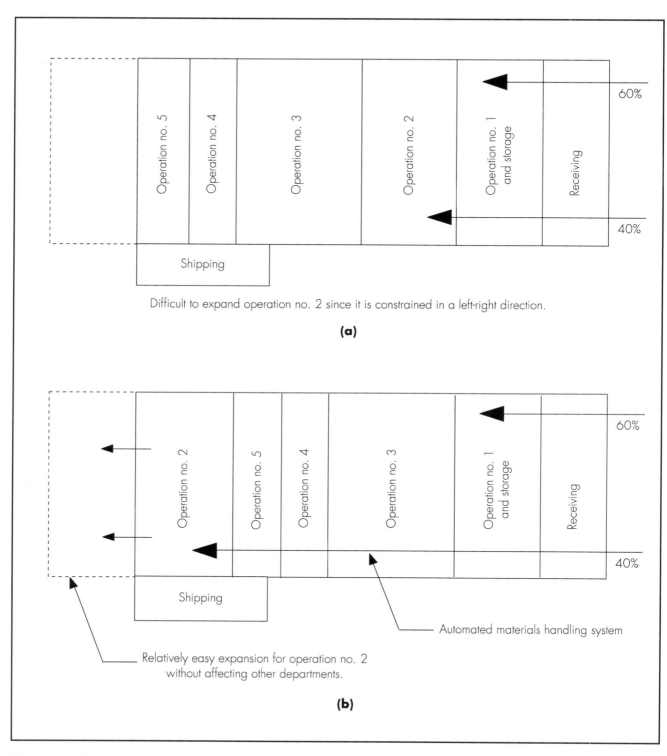

Difficult to expand operation no. 2 since it is constrained in a left-right direction.

(a)

Automated materials handling system

Relatively easy expansion for operation no. 2
without affecting other departments.

(b)

Figure 15-1. Though closeness and activity relationship analyses can be quite helpful in planning layouts, they must be integrated with materials handling choices to prevent an inefficient design approach. In the case shown here, if only relationship analysis were used, Operation 2 would stay where it is sequentially (a), and result in a difficult expansion task.

The basic factors affecting all moves are shown in Table 15-1.

The benefits of a well-designed materials handling system include:

- Generates savings in ongoing materials handling transport costs;
- Helps to force a straight-through, "handle once," single-container usage approach;
- Helps to force a standardized container approach that simplifies materials handling equipment choices and can provide the impetus for total plant materials handling mechanization;
- Can help lower inventory and production control costs;
- Can help to provide better space utilization;
- Can increase production capacity;
- Can help to implement a shorter production time cycle;
- Can reduce parts damage and waste;
- Can provide a high degree of safety;
- Can improve a company's image with its customers;
- Can help to improve employee morale.

Table 15-1. Basic Factors Affecting All Moves

1. Materials
    –Physical characteristics
    –Other production-related characteristics

2. Routings
    –Distance
    –Physical situation

3. Intensity of flow

4. Condition of flow (movement)
    –Quantity conditions (regularity, etc.)
    –Service conditions (separate lot heats, etc.)
    –Timing conditions (speed, synchronization)

Straightening out and streamlining materials flow in and of itself would be a great benefit to any company, whether it is using manual or mechanized materials handling approaches.

A discussion of materials handling equipment would not be complete if we did not also consider the negatives as well as the positives associated with some fixed equipment choices (e.g., powered conveyors, drag lines, monorails, cranes, automatic storage and retrieval systems, etc.). Two main negatives are inherent in these types of mechanized approaches: the initial capital investment requirement and the lack of full flexibility to meet changing conditions. If a company expects to have continuing major changes in its parts, products, or processes, then *flexibility* should be one of the most important factors to consider. For example, conveyor systems can be moved and rearranged rather quickly, but they do tend to place "blinders" on layout planners. Planners will frequently try to work around the installations rather than redesign or move them. This sometimes prevents planners from reaching optimum solutions. Likewise, floor drag lines typically are sequenced systems which are not that expensive to move but are frequently considered immovable monuments by planners.

Monorail systems, unless they are of the power and free type, can also inhibit full flexibility of materials movement. They tend to force serial, sequenced, deliveries of materials where a missed "pickoff" can have serious consequences. Frequently, a missed pickoff of a particular item on a drag line or monorail system will force the material to travel the entire length of the loop before returning to the correct pickoff point, similar to an airport baggage conveyor. Since these systems tend to be very long, the complete loop travel time is usually very long. A manual contingency method must be in place to handle these events.

APPENDIX

SPACE PLANNING SOFTWARE AVAILABLE IN THE U.S.

A Sampling

Vendor	Product	Telephone
CIMtechnologies Ames, Iowa	FACTORY OPT™ and FACTORY PLAN™	515-296-9914
Production Modeling Dearborn, Michigan	LAYOPT™	313-441-4460
Archibus, Inc. Boston, Massachusetts	ARCHIBUS/FM™	800-541-2724
SABA Solutions Manhattan Beach, California	WinSABA™	310-379-9136
CAFM Works, Inc. Cambridge, Massachusetts	CAFM Space/Stack and Block™	617-492-1148
High Performance Systems, Inc. Hanover, New Hampshire	STELLA II™	603-643-9636
Innovative Tech Systems Inc. Horsham, Pennsylvania	SPAN™	215-441-5600
Cadapult Ltd. Newark, Delaware	CAFM-space Analysis™	302-594-9416
University of Houston (Texas) Industrial Engineering Dept.	BLOCplan™	713-221-8000

INDEX

A

Activities listing, 180
Adjusting today's needs, 106
Affinities (see also relationships), 135
Aisle(s)
  bridging the cross-, 113
  efficient *a.* grid, 174
  main, 21, 23, 95
    space, 96, 97
  planning natural *a.* patterns, 172
  straight-running, 172
    cost advantage of, 172
  typical widths, 110
Alternative layout configurations, 175
  logic and attributes, 175
    typical factors, 175
Alternative sites
  evaluating, 16
Apple, James M. Sr., 65
Area(s)
  allowances, 23
  aspect ratios, 21
  "nonproduction," 139
  requirements, 180
  shapes, 21
  transaction-related, 164
Attributes
  defining, 40
  success, 40
  typical layout, 40
Automatic storage and retrieval
  system (ASRS), 113, 117, 118
Automation, 85
Awkward load, 87

B

Backflows, 92
Basic questions, 9
Block layout(s)
  final, 183

Bottleneck
  position, 83
  sensitivity analysis, 148
Budget limitations, 190
Buffer(s), 84
  inventory, 105
  storage, 80
  temporary, 80
  zones, 84
Building constraints, 129
Business
  focused *b.* units, 6
  trends, 99

C

Capacity, 80
  utilization, 125
Carousel(s)
  horizontal, 113, 119
  vertical, 113, 120
Centroidal(s)
  area points, 21, 22
Charting process, 144
  combined flow, 144
  other-than-flow, 144
Choke point, 83
Chrysler, 85
Circulation conveyor, 86
Classic relationship chart, 140
Classification procedures
  refining, 144
Clear
  equipment *c.* heights, 34
  heights, 171
  span, 14
    construction, 180
"Closeness," 134, 138
  priorities, 178
Coding
  color, 141
  schemes, benefit, 200

Common
  denominator handling unit, 68
  parts, 85
Competitive
  advantage, 1
    one-time, 37
  business challenges, 1
Computer
  analysis
    macro flows and costs, 231
  assembly, 93
  -based tools, 231
  dynamic model building, 236
  dynamic simulation, 234
  simulation packages, 235
Constraints, 166
  budget, 165
Container(s)
  configuration, 67
  knockdown, 68
  nest(able), 10, 68
  return flow of empty, 79
  returnable-type, 68
  shipping, 68
  size, 67
  type(s) of, 64
    collapsible steel, 64
    collapsible wire, 64
    corrugated metal, 64
    returnable, 65
    shop box, 64
    shop truck on casters, 64
    stacking tote boxes, 64
    wirebound wooden pallet, 64
    wooden box on casters, 64
    wooden pallet box, 64
  universal, 68
Continuous improvement doctrine, 2, 3
Conversion method, 106
Copyright laws, 129
Cost(s)
  handling, 87

materials handling, 192
materials transport, 161
minimizing materials handling, 90
-related subjects, 93
traditional items, 196
Crane(s)
overhead, 172
pick-and-place, 121
Cube
between activities, 88
movement(s), 88
utilization, 120
Cycle time(s), 3, 68
lot *c.t.* , 67

D

Data
basic *d.* needs, 43
collection, 41
fundamental, 45
issues and problems, 46
degree of mechanization/
automation, 48
degree of specificity, 47
establishing optimal relationships, 47
estimating flexibility required, 47
resolving accuracy problems, 46
weighting factors in evaluation, 47
review, 43
four major facets, 43
Decision support system (DSS), 40
Deming, 80
Depackaging, 10
Detail layout(s)
fundamental questions, 185
Detrashing, 10, 79
Discrete event
approach, 2
philosophy, 3
Drawings
"as-built," 25
Dual manufacturing, 68

E

Employee(s)
cross-trained, 122
long *e.* walks, 5
projections, 95
Entrance/exit points, 70
Environmental assessments, 17
storm water collection, 17
topography, 18
Equipment
low utilization, 200
utilization considerations, 105, 106
Equivalency
establishing, 87
factor, 87

"Equivalent"
containers, 87
pallet load, 87
unit loads, 90
Ergonomics/human factors, 186
anthropomorphic data, 187
empirical "rules," 187
equipment aids, 189, 190
"sweet" zone, 186
Excavation
under footing, 15
Expansion
crane bay, 12
flexibility considerations, 176
plant, 38
probability, 149, 177, 184
intensity, 150
sensitivity analysis, 176
Experience curve, 125

F

Factors affecting all moves, 251
materials, 92
methods of handling, 92
money, 92
moves, transport distance, 92
First lot output, 68
Flexibility, 21, 48, 148
degree required, 48
for capacity expansion, 40
of physical change, 40
Flexible manufacturing systems, 199
Flexible overcapacity, 105
Floor(s)
loading restrictions, 172
observations, 92
super flat, 11
Flow(s)
below the diagonal, 92
intensity(ies), 134, 137
diagramming, 138
measurements, 85
-of-materials effectiveness, 40
volume, 85
Flow patterns, 70
frequently overlooked, 79
materials, 71
typical manufacturing, 69
Focused-factory strategy, 130
Footprint
diagram imposed on, 181
/inventory reduction, 113
Ford, 84
From-to chart, 88, 89, 91

G

Gantt charts, 31
Group technology, 199

H

Handling
ease, 87
increment, 68
Harley-Davidson motorcycle assem-
bly, 120
Hash mark count, 90
Heat-treat, 67
"Humanistic" system, 84

I

Industrial
long *i.* truck movements, 5
Intradepartment/cell layout considerations, 180
Inventory(ies),
critical care, 116
importance of *I.* turns, 124
stocking objective, 124
turnover goals, 43

J

Just in time (JIT), 164
approach, 59

K

Kanban
considerations, 122
containers, 123
size, 123, 124
detail counting, 123
usage, 124
system, 123

L

Labor balance, 82
Large load principle, 66
Logic approvals, 150

M

Mag count, 85
Manufacturing
assembly line
balancing labor, 219, 220
cell(s), 59, 197
analysis and optimization, 209
balancing labor, 219, 220
characteristics, 197
classification by process, 213,
214, 215
commonality characteristics, 204
design questions, 208
detail grouping, 208
detail planning, 210

history, 197
interest in, 198
macro-grouping factors, 204
methods improvements, 210
people issues, 201
planning procedure, 204
procedural issues, 201
process issues, 203
process per class, 215
roles/organization, 207
key fundamentals in, 5
"lean," 164
push versus pull, 59
cons of pull system, 60
pros of pull system, 60
requirements/resource planning
(MRP) systems, 59
spectrum, 49
high-volume, low variety
production, 49
low-volume, high variety
production, 49
variety, 199
Market(ing)
focus, 208
long-term projections, 126
Master plans, 101
long-term site, 193
Material(s) flow
analysis, 41, 135
costs, 192
"intensity," 85, 90, 131, 191
mode of, 161
non-homogeneous, 85
seasonality effects, 124
sequence, 90
work simplification, 65
Materials handling
analysis, 63
basic questions, 63
benefits, 249, 251
bulk versus unit, 63
dead heading, 65
discrete handling containers, 63
discrete unit loads, 63
equipment choices, 249
homogeneous, 90
KISS system, 63
minimizing, 63
nonvalue-added, 65
total costs, 63
Mezzanine areas, 172
Miniload storage and retrieval
systems, 113
Monuments, 19, 20, 45, 131, 166, 167
reverse, 45, 46
Movement of materials
generic, 92
Multiproduct chart, 92
Muther, Richard, 85

N

"Natural" directional sequence, 88
Negative impacts, 2
Normal
curve, 83
distribution, 82

O

O, P, Q, R, S, T approach, 44
Operations
clone current *o.*, 37
process (flow) chart (OPC), 71,
73-75, 135, 177
consolidated *o.*, 76- 78, 93
multiproduct, 71, 76
sequential, 81
receiving, 71
reducing nonvalue-added, 203
shipping, 71

P

"Paralysis by analysis" effect, 130
Parking areas, 101
Part
classification, 203
/product classification, 211
Pattern of procedures, 41
People
skills and leadership, 39
10 major attributes, 39
centralization of technical, 61
Phases
major planning, 8, 20
overlapping, 8
Piecepart
batch-quantity, departmental
approach, 59
single *p.* flow, 59
Plan(s)
master facility, 101, 103
site saturation, 101
Planning
block layout, 19
bottom-up approach, 25
computer-aided approaches, 157
ALDEP™, 158, 160
CORELAP™, 158, 159
CRAFT™, 158
space planning (CASP), 157
conventions and standard symbols, 42
detail layout, 25, 26
equipment and systems, 9
initial fundamentals, 9
systems philosophies, 9
implementation, 27

layout *p.* approach, 41
manufacturing
cell, 206, 207
process, 8
natural aisle patterns, 172
overall site, 18
space, 10
buffer loop, 12, 13
generalities, 11
industrial, 11
joint team effort, 11
major aisle, 16
most common error, 11
parking areas, 16
receiving, 14
shipping, 14
site questions, 16
storage areas, 109
top-down approach, 25
Plant layout(s)
batch-oriented, 55
-by-process, 58
combination batch and cell, 56
evaluating alternatives, 245
decision factors, 245
five phases of developing, 7
block layout, 7, 20
detail layout, 7, 20
installation, 7, 20
location analysis, 7, 20
needs analysis, 7, 20
high-aspect-ratio configuration, 169
macro, 90
marriage of *p.l.* and materials
handling, 10
part-time efforts, 165
product-focused, 57
product-oriented, 54
rough-cut macro-block alternatives, 90
the three *As* , 133
Adjustment, 133
Affinities, 133
Areas, 133
Power columns
vertical, 14, 113, 120
Powered takeaway line, 86
Precedence diagram(s), 220
Ranked Positional Weighted (RPW)
method, 221
"Principles of Material Handling," 65
Process
centralized *p.* layout, 59
focused philosophy, 55
versus product, 54
Product/Quantity
Chart, 50, 54
combining operations, 51
curves, 54
direct labor per part, 51

gross profit contribution, 51
macro planning, 50, 54
potential splits and combines, 51
quantity, 51
splitting operations, 51
Product versus quantity, 50
Production line
  and cell balances, 79, 83
    importance, 79
    opportunities, 79
  sequence-dependent, 82
  slowing, 4
Production system
  "lean," 84
Project management
  computer-based, 28
  tool, 39
Pull system
  replenishment, 123, 124
Push basis, 123

Q

Quality
  product, 6

R

Rack(s)
  double-deep installation, 110, 114
  nestable, 69
  stackable, 69
Rating(s), 136
  reasons for, 141
Ratio(s)
  business activity-dependent, 98
  common, 125
  gross business, 98
    use of, 95
  manufacturing space, 98
  space and revenue, 95
  trend, 100
    and projections, 125
Recirculating
  conveyor, 85
  lines, 86
Relationship/affinity analysis, 129
  engineered planning approach, 129
    clone method, 129
    committee method, 129
    consultant's method, 131
    straight-materials-flow
      method, 130
    strong-man method, 130
Relationship(s)
  activity(ies), 22
    establishing r. between, 134
  based on materials flow, 135
  based on nonflow factors, 139
    site-specific, 148
  chart, 142, 143, 178

combined, 148
combining, 146
diagram(s), 151, 155, 157, 179
  affinity, 151
  benefits and pitfalls, 160
  bubble, 151, 152
  for existing plants, 163, 171
  for "greenfield" site, 169
  forging, 193
  new plant versus expansion, 163
  physical considerations, 171
  pre-diagramming, 163
  proximity, 151
  spatial, 169, 170, 182
diagramming, 138, 153
flow
  combining and weighting, 144
intensity, 154
nonflow
  combining and weighting, 144
pairs, 144
total *tie*, 135
Return on investment (ROI)
  short-term, 2
Revenue
  annual *r.* projections, 99
  per employee, 100
"Rough estimate" case, 106
Routing charts, 80

S

Sales
  revenue projection, 98
Sampling, 92
Scheduled push
  systems, 59
    versus demand pull, 59
Schematic
  energy planning, 24
    compressed air, 24
    power demand factor, 24
  utilities, 24
SECS, 65
Seismic probability, 27
Side loader
  double-reach, 115
Single-part flow, 67, 84
Site saturation, 103
Society of Manufacturing Engineers
  (SME), 38
Space
  allocations,
    multifloor, multisite, 225
  balance, 102
  balance analysis, 101
  balance for long-term projections, 98
  computer allocation systems, 226
  detail determination of needs, 109
    rough layout method, 109

differential comparison, 106
pitfalls of *s.* projections, 127
planning software, *(appendix)*
planning worksheet, 107
productivity, 98
projections, 104
requirement(s), 95
  calculating, 95
  overstate, 95
underroof *s.* utilization, 102
usage record, 101
versus picking and replenishment
  labor, 125
"Spine" approach, 71
Standard time(s), 80, 81
  labor, 83
Storage,
  modules, 110
  of pallet loads, 110
  over-the-aisle, 116
  "pick" access considerations, 212
  point-of-use philosophy, 116
  shelf, 122
  size of *s.* areas, 124
  typical module layouts, 111, 112
Supplier(s),
  certified, 71, 93
Systematic Layout Planning (SLP), 85

T

The Goal, 84
Timing diagram(s), 220
Toyota, 84, 85
  approach, 123
Transaction
  intensity of, 88
Transport
  costs, 66
  horizontal inplant, 10
Trend analysis, 126
Truck(s)
  aprons, 19
  direction, 19
  dock configuration, 18
  road width, 19
  turret, 115
Typical part's life, 205

U

Unit load(s)
  effects of, 67
  equivalent *u. l.* analysis, 85
  large versus small, 65
Unit size principle, 65
Utilities
  power, 172
  sewer, 172
  water, 172

V

Value-added assessment, 204
Visual replenishment systems, 123
Volatile organic compounds (VOCs), 32
Volume, 85
　cubic, 87
　　of materials, 90
Volvo, 84
　method, 84
Vowel/letter conventions, 138

W

Weighted factor
　analysis, 247
　evaluation, 245
Weighting scheme, 145
　pitfall, 145
Work elements, 217
Work in process (WIP)
　backflushed, 123
　breeding, 58
　inventories, 43
　　minimum, 123
　reductions in *w.* inventories